Living in a Dangerous Climate

Living in a Dangerous Climate provides a journey through human and Earth history, showing how a changing climate has affected human evolution and society. Is it possible for humanity to evolve quickly, or is slow, gradual, genetic evolution the only way we change? Why did all other *Homo* species go extinct while *Homo sapiens* became dominant? How did agriculture, domestication, and the use of fossil fuels affect humanity's growing dominance? Do today's dominant societies – devoted as they are to Darwinism and survival of the fittest – contribute to our current failure to meet the hazards of a dangerous climate? Unique and thought provoking, this book links scientific knowledge and perspectives of evolution, climate change, and economics in a way that is accessible and exciting for the general reader. This book is also valuable for courses on climate change, human evolution, environmental science, and environmental economics.

Renée Hetherington obtained a BA in business and economics from Simon Fraser University in 1981; an MBA from the University of Western Ontario in 1985; and an interdisciplinary PhD in anthropology, biology, geography, and geology from the University of Victoria, British Columbia, in 2002. She was awarded a Canadian National Science and Engineering Research doctoral fellowship for her work reconstructing the paleogeography and paleoenvironment of the Queen Charlotte Islands/Haida Gwaii. The Canadian Social Sciences and Humanities Research Council subsequently awarded her a postdoctoral fellowship for her research relating climate change to human evolution and adaptability over the last 135,000 years. She has been coleader of International Geological Correlation Program project 526, "Risks, Resources, and Record of the Past on the Continental Shelf." She and her husband Bob are partners in RITM Corp., a consulting company committed to helping organizations, especially in the resource sector, reach their potential while recognizing that we live in a changing world. She ran for office as a Member of the Canadian Parliament in 2011 and is currently a member of Shadow Caucus with the Federal Liberal Party of Canada. Renée is the coauthor (with Robert Reid) of *The Climate Connection: Climate Change and Modern Human Evolution* (Cambridge University Press, 2010). She lives with her family on Vancouver Island.

"… very important and valuable book that will help people realize how dangerous times are and why they are dangerous. [It explains] what we as humans are doing to make the changes more destructive … [and how] our evolutionary perspective as a species ordains us to want to continue our self-destructive ways."

– Fred Roots, Chair of the Canadian National Committee for the UNESCO Man and Biosphere Programme

"Renée Hetherington marshals evidence from anthropology, biology, and earth science to discuss the imminent fate of the human species. She argues that we are our worst enemy, caught in the mental gridlock of a neo-Darwinist, market-based capitalist system that is preventing us from dealing with impending serious climate change and the possible collapse of the ecosystems that support us. Yet Hetherington offers hope – *Homo sapiens* is the most adaptable of animals; we can, if we choose, deal with the most serious challenge our species has faced. Hetherington challenges us to act. Failure to meet the challenge would be the ultimate irony for the most intelligent life form that has ever graced Earth."

– John J. Clague, Director, Centre for Natural Hazard Research, Simon Fraser University, Canada

"This book addresses the history of climate variability and recent anthropogenic climate change on the background of human evolution; discusses the links between climate, migration, and agriculture; and assesses the dominant paradigm, the economic connection, dangerous attitudes, and the main new dangers due to climate change impacts. Written from an interdisciplinary perspective, this important and well-written scientific book deserves a wide public audience of students, citizens, and opinion leaders in society, business, and politics to counter the well-financed campaign of economic lobbyists and climate skeptics who ignore the impact of non-action on future generations."

– Ursula Oswald Spring, National Autonomous University of Mexico; lead author of the IPCC

"A wide-ranging and thought-provoking exposition that explains the human condition, from our evolution to our inability to deal with current, let alone future, climate change in the face of an imperfect financial system. Dr. Hetherington is a skilled storyteller who provides a factual yet, at times, personal account that integrates several fields of endeavor and leaves worrying questions about how we will cope with the challenges of the future. A highly readable and understandable synthesis that should have wide appeal."

– Allan R. Chivas, GeoQuEST Research Centre, School of Earth and Environmental Sciences, University of Wollongong, Australia

Living in a Dangerous Climate

Climate Change and Human Evolution

RENÉE HETHERINGTON
RITM Corporation

CAMBRIDGE UNIVERSITY PRESS
Cambridge, New York, Melbourne, Madrid, Cape Town,
Singapore, São Paulo, Delhi, Mexico City

Cambridge University Press
32 Avenue of the Americas, New York, NY 10013-2473, USA

www.cambridge.org
Information on this title: www.cambridge.org/9781107694736

First published 2012

Printed in the United States of America

A catalog record for this publication is available from the British Library.

ISBN 978-1-107-01725-2 Hardback
ISBN 978-1-107-69473-6 Paperback

Contents

	List of Figures	*page* vii
	List of Tables	viii
	Foreword: Evolution and the Human Condition *Robert G. B. Reid*	ix
	Acknowledgments	xiii
	Preface	xv
	EARTH'S CLIMATE: IMPACTS ON HABITAT AND HUMANS	
1.	Putting Our Emergent House in Order	3
	THE EVOLUTION OF THE *HOMO* SPECIES	
2.	The Cradle of Humankind	23
3.	The Neanderthal Enigma	36
4.	The End of *Homo* Diversity	41
	CLIMATE AND HUMAN MIGRATION	
5.	Climate and Human Migration	47
6.	Braving the New World	58
	CLIMATE AND AGRICULTURE	
7.	Agriculture and the Rise of Civilization	73
8.	The Maya Civilization and Beyond	88

THE DOMINANT PARADIGM

9. Dominance Destabilized 99
10. Fitness Folly 106
11. Darwin the Selector 112
12. Hunting Down Woody 116
13. Kammerer's Suicide 119
14. Giants and Pygmies 123
15. Dutch Hunger Winter Babies 127

TODAY AND TOMORROW

16. Today and Tomorrow 133
17. Dead Zones 146

THE ECONOMIC CONNECTION

18. The Economic Connection 155
19. The Progress of Dominance 168

DANGEROUS ATTITUDES

20. Dangerous Attitudes 175
21. Helpful Strangers 180
22. Triumphant Oblivion 184

LIVING IN DANGEROUS TIMES

23. Our Children 191
24. Living in a Dangerous Climate 196

Glossary 209
Notes 227
Index 247

Figures

1.1.	Geologic timescale	*page* 5
1.2.	Quaternary timeline	8
1.3.	CO_2 records and temperature reconstruction of lower atmosphere over the last 800,000 years based on air trapped in ice cores from Antarctica	10
1.4.	Global carbon dioxide emissions from human activities, 1750–2004	14
1.5.	A composite atmospheric CO_2 record over 650,000 years – 6.5 ice-age cycles – based on a combination of CO_2 data from three Antarctic ice cores: Dome C, Vostok, and Taylor Dome	15
2.1.	Key early hominin sites in Africa	26
2.2.	The fall of Neanderthals and the rise of modern humans	33
5.1.	The extent of glacial ice covering the continents during MIS6	48
5.2.	The relative sea-level curve for the last glacial cycle	49
5.3.	Hypothetical routes out of Africa, highlighting a southern coastal route along the Indian Ocean coastline that may have been taken by humans migrating out of Africa	53
6.1.	Land elevation and ocean depths showing continental shelves of the world and indicating the location of the Bering Strait continental shelf (Beringia)	60
8.1.	Sculptures from Maya city of Cobá, Yucatan Peninsula, Mexico	89
8.2.	Nohoch-Mul, Maya	90
8.3.	View of northern Yucatan from Nohoch-Mul	91

Tables

2.1. Hominin species *page* 27
7.1. The timing of plant and animal domestication; points of origin of domesticated species 80
16.1. Global effects of rising temperatures and atmospheric CO_2 concentration levels 140

Foreword: Evolution and the Human Condition

ROBERT G. B. REID

While the twentieth century stands out in history for two world wars, numerous local wars, genocides, and political revolutions, there were little-considered developments that present us in the twenty-first century with even more menace. Because they have crept up on us slowly, instead of with the instant and obvious catastrophic effects of war and revolution, most of us tend not to notice them or to discount their effects. I refer to the environmental consequences of global warming, deforestation, soil erosion, the degradation of ocean fisheries, expanding populations, and inflated economies.

Renée Hetherington confronts us with these present, if not clear, dangers, as well as with the risks of living beyond our means. She ranges from an intimate subjective point of view to a scholarly analysis backed with solid evidence. Her two-pronged approach to these problems arises from her knowledge of climatological history and from her belief that because economics and business are devoted to Darwinism and survival of the fittest, they are partly to blame for the failure to meet the growing hazards that are going to make the world a more dismal place in which to live.

Following her argument that the dynamic stabilities of biological, political, and economic systems, which fit into the Darwinist mold, tend to resist progressive change, she points out that they can only be disequilibrated by unpredictable natural catastrophe or by objective analysis and resolute action on the part of humankind. She tells us how natural disasters shaped the future of evolution, and she provides the objective analysis required for resolute action.

As an evolutionist, I appreciate how Hetherington points out some of the flaws of current evolutionary theory based on natural selection, suggesting that some biological change can occur nonrandomly, regardless of adaptiveness. She argues that some biological change can be interpreted in terms of the Lamarckian concept of "the inheritance of acquired characteristics." She also reasons that some evolution can be directly driven by the environment.

I applaud the distinction she makes between adaptation and adaptability. The two are hopelessly confused in the public mind as well as in the writings of anthropologists and archeologists. The *adaptations* with which neo-Darwinists work are random, genetically fixed mutations that require the approval of natural selection to become general species characteristics. *Adaptability* is what the individual organism can do to respond physiologically and behaviorally to change. In the case of humans, the proper application of intelligence is part of our adaptability. The distinction is particularly important in the context of this book, because, as Hetherington points out, the process of adaptation and natural selection in the strict sense is much too slow to respond to sudden environmental alteration. In contrast, the adaptable organism can do something about it instantly. Unfortunately for humans, tradition, ritual, and "sticking to tried and true ways" – what Renée Hetherington calls "the dominant attitude" – can obstruct effective action, despite our potential ability to respond effectively to change.

A prevalent theme of this book is the concept of the "hopeful monster," proposed by Richard Goldschmidt in 1933. He believed that some kinds of mutation could produce radically new plants and animals. They "hoped" for an ability to reproduce their novel qualities and for conditions to which they could be adaptive. The hopeful-monster hypothesis was rejected by neo-Darwinist gradualists who thought that every change was small and judged fit by natural selection before being added to the genomic repertoire. As Hetherington points out, Darwin was well aware of "monstrosities" that arose suddenly during animal breeding. Although he used artificial selection (i.e., what farmers do when they decide particular plants and animals are more suitable for their purposes) as a model for his evolutionary theory, he rejected natural monstrosities as aberrations because they did not fit in with his concept of gradual evolution by natural selection.

A monstrosity that is unfit in the Darwinian sense can nevertheless survive if a farmer or horticulturist provides appropriate conditions. Likewise, in nature, unfit monsters can get by if they can find such conditions for themselves. If the monsters' hopes are realized, they become "successful monsters"; their populations, environmental distribution, and diversity expand, and they become registered in the fossil record. The evolution of our vertebrate ancestors provides us with numerous examples of the episodic, sudden changes that Goldschmidt envisioned, albeit few and far between. However, as successful monsters, we humans have periodically mapped roads to our own downfall without taking significant action to forestall such fates. In the case of the present danger, it will affect the entire population of Earth.

In writing this book, Renée Hetherington mobilizes her interdisciplinary background in business administration, anthropology, climatology, and biology. It is a fresh look at a subject that has been tackled with little success by many others – a new way of seeing. In addition to her scientific knowledge, she draws on personal experience, sometimes poignant, sometimes entertaining. She writes in plain language rather than scientific jargon and makes her work accessible to most readers.

Acknowledgments

This book grew from my desire to make my and Robert G. B. Reid's recent book, *The Climate Connection: Climate Change and Modern Human Evolution*, accessible to a broader audience. For years, Robert and I met weekly over lunch to discuss evolution, the environment, and humans. His support and encouragement never wavered, even after he became ill. I deeply appreciate all that he has done to assist and encourage the success of my work, including writing the foreword for this book.

Our discussions frequently benefited from the contributions of Richard Ring and numerous other faculty and students from the University of Victoria biology department, who joined us in our discussions. Researchers in Andrew Weaver's University of Victoria climate modeling lab gave me the opportunity to stray into unchartered territory, and with the support of the staff – particularly Michael Eby, Wanda Lewis, and Ed Wiebe, postdoctoral fellows at the time –Jeff Lewis and Kirsten Zickfeld, and students, I gathered the research that contributed to the writing of this book. My appreciation also extends to Tina Sherlock, public services librarian at Quest University, who graciously provided me access to library resources.

The many colleagues, public citizens, and students with whom I have spoken and to whom I have lectured at elementary schools, universities, and scientific institutes asked questions that provoked me to search for clear, insightful responses and helped me focus on the critical issues.

E. Fred Roots, science advisor emeritus with Environment Canada, dug up and presented me with climate policy research papers and notes extending back more than twenty-five years, providing me with perspective on the current climate debate.

My deepest appreciation goes to my editor, Matt Lloyd, and the rest of the team at Cambridge University Press, who provided excellent advice and assistance. Matt believed in this book long before it was written; he gave me the encouragement and flexibility to write it and the opportunity to publish it.

My sincere gratitude is extended to Audrey McClellan, my personal editor, who has worked with me and this book over several years, thoughtfully and professionally guiding me to my objective of making scientific knowledge and ideas accessible to the general public. My appreciation also extends to Bhavani Ganesh and the Newgen Knowledge Works production team, as well as to Fred Goykhman of PETT Fox Inc. for his copyediting and editorial assistance.

Adam Monahan and Robert G. B. Reid provided thoughtful and insightful scientific reviews of earlier versions of this book, which have improved its clarity and content.

I sincerely appreciate the insights provided by John Tapping, who reviewed and provided contributions for the "Capitalism and Democracy" chapter and the assistance of Ryley Tapping and the Hernandez family for providing access and insights to Maya culture in Mexico.

The complexities of my father's life stimulated some of the personal anecdotes and thoughts on survival of the fittest, and those of my mother's life drew me to ponder the capacity and benefits of human compassion.

My dear friend Marion Farrant offered me encouragement, advice, and support from the outset and reviewed numerous versions of this book; without her, this book would never have been written.

My sons, John and Ryley, provided me opportunities to write and do research. They gave willingly of their time and support, love, and compassion; they also give me hope and optimism for the future of humanity.

My husband, Robert, the love of my life, has never wavered in his support and love. He personifies my compassionate and tolerant theoretical jargon.

Any errors, omissions, or inconsistencies are solely my responsibility and in no way reflect on the contributions made by all the people I have mentioned and the many I have not named here but appreciate nonetheless.

Preface

Given things as they are, how shall one individual live?

Annie Dillard, *For the Time Being*

This is a book about humans, the rise of our civilization, and the development of our economic system. It is about our relationship with Earth, the species with whom we share this planet, and the climate that influences Earth's environment. It covers a diverse range of subjects; its relevance is applicable across disciplines and cultures; its implications are vital. I have tried to convey the scientific evidence, which is frequently buried in research journals inaccessible to the general population, in language that is as nonscientific as possible in an effort to make clear to the reader the brutally pressing reality that what matters most is how each of us lives our life, particularly as it relates to climate, our economy, and the increasing dominance of humans on Earth.

Charles Darwin's theories of survival of the fittest and natural selection form the dominant paradigm against which recurring new species – "hopeful" monsters – perplexingly arise and, sometimes, survive. Others, like all previous *Homo* species described in Chapters 2 through 4, did not survive. Individual examples described in Chapters 11 through 15, some more hopeful than others, serve as opportunities to review our perspective of Darwin's theories, which dictate that humanity's burgeoning presence represents expanding wisdom and fitness, yet simultaneously feed an omnipresent denial of death and destruction.

Chapters 5 and 6 take the reader on a historic journey with our ancestors as they migrate from Africa to Eurasia and then on to the Americas. Early efforts to control the effects of a changing climate through agriculture and early civilization are revealed in Chapters 7 and 8. The distinction between adaptation and adaptability and how dominance prevails during climate stability, whereas flexibility is critical during rapid climate change, are discussed in Chapter 9. Chapter 10 delves into the wonders and follies of natural selection and survival of the fittest. Given these connections between humans and climate, Chapters 16 and 17 review what the environment holds for humanity today and tomorrow.

I discuss the similarities between the underpinnings of economic theory and evolutionary theory in Chapters 18 and 19 and why economic and societal dominance cannot be guaranteed during periods of significant change and instability. I also remark on the impact of human dominance in a world where climate change is not new but is part of the natural change ubiquitous in Earth's history, pointing out that climate change has often coincided with the extinction of once-dominant species.

In the closing chapters, I reflect on our place on Earth today, on our capacity to effect change, and on our ability to live in a dangerous climate. Storm clouds brew on the horizon. In the past, we rarely worried about our footprint – a wave would always come and wash it away. Now, increasingly, the waves fail to clean up our mess, or they come as tsunamis, obliterating everything in their wake.

To a considerable extent, humans remain oblivious to the consequences of dominance, perhaps because we still survive or, alternatively, because we are ever hopeful. Yet, irrespective of *Homo sapiens*' capacity to intellectually and behaviorally adjust to our new reality, Earth does not have a preference for a dominant species. It does not care which rules the day. Change is its wardrobe.

Earth's Climate: Impacts on Habitat and Humans

1

Putting Our Emergent House in Order

Nothing in the world lasts
Save eternal change.

Honorat de Bueil, seigneur de Racan (1589–1670),
"The Coming of Spring"

Dry winds blow across drought-ridden, rain-parched farms in Australia, followed by record flooding. Heavy rains inundate low-lying deltas in Asia and on the Pacific northwest coast of North America, flooding homes and making water unfit to drink. Death comes to pastoralists and farmers on the shrinking acreage of arable land in Africa as too little rain feeds too few crops. Intense hurricanes batter southern Atlantic coastal regions. The world looks on as Africa suffers escalating destruction; rising prices and shortages of food and basic goods, along with joblessness and stagnant wages, trigger protests and the destabilization of political regimes in Africa and the Middle East. Riots bring London to its knees. A global recession cripples the world as government debt soars in Europe and America. People die. At global summits on climate change, politicians refuse to sign agreements that would have them reduce their countries' emissions of the greenhouse gases blamed for causing global warming, or they renege on their previous emission-reduction commitments. Media reports swing between predictions of climate catastrophe and such derisive statements as "If we can't predict the weather next week, how can we predict climate next year?" A perplexed public is left not knowing whom to believe.

Most of us have unanswered questions about the real state of Earth's climate and its impact on local and global economies. This lack of understanding makes us feel helpless and uncertain about what we need to do and where we fit in this changing world. Assembling answers to some of those questions will help us look at the issue more clearly, without feelings of panic or hopelessness.

Is climate change a natural phenomenon on Earth?

Yes, it is. The Earth is 4.5 billion years old, and before the first signs of life appeared, the land and ocean formations on Earth looked vastly different from what we see today. Even the continents situated so solidly beneath our feet were shaped and positioned differently. There was little or no oxygen in Earth's atmosphere then, hence no ozone and no protection from solar radiation. Initially, Earth's atmosphere contained hydrogen and helium. Later, as the planet cooled, it sent gases, including carbon dioxide, water vapor, and possibly methane into the atmosphere. Single-celled microbes released oxygen, a by-product of photosynthesis, which gradually accumulated in the oceans and atmosphere.

Less than a billion years ago, nearly all the surface water on Earth froze, and most of the planet was covered in ice, creating what is known as "Snowball Earth." Over millions of years, large volcanic eruptions emitted enough heat and carbon dioxide into the atmosphere to create a greenhouse warming effect that melted the ice and brought an end to the "snowball" state. These dramatic fluctuations in climate had a devastating effect on the life forms evolving in the oceans, but about 500 million years ago (for a timescale of the geological history of Earth see Figure 1.1), the big bang of animal evolution took place. This saw the origin of all the major phyla (groups of organisms). By about 450 million years ago, plants had invaded the land, their vegetation covering the previously barren landscape. And 50 million years ago, Earth's climate was sufficiently warm for the Arctic to experience a Mediterranean climate with temperatures as high as 24°C.[1] And the changes continue.

Does climate change cause species extinctions?

History suggests it does. Five major extinction events have occurred on Earth in the past 450 million years, and scientists believe that at least one of them – which took place at the end of the Cretaceous period 65 million years ago – may have been related to sudden and catastrophic climatic changes caused by bolide impacts (i.e., meteorites, comets, or asteroids hitting Earth) or other geological events, including large volcanic eruptions that spewed forth huge lava flows.[2]

Eon	Era	Period	Epoch	Age in millions of years
Phanerozoic	Cenozoic	Quaternary	Holocene	0.01
		Quaternary	Pleistocene	2.6
		Tertiary: Neogene		23
		Tertiary: Paleogene	Oligocene Epoch	34
		Tertiary: Paleogene	Eocene Epoch	56
		Tertiary: Paleogene	Paleocene Epoch	65
	Mesozoic	Cretaceous		146
		Jurassic		200
		Triassic		250
	Paleozoic	Permian		299
		Carboniferous		360
		Devonian		416
		Silurian		444
		Ordovician		488
		Cambrian		542
Precambrian	Proterozoic	Neoproterozoic		1,000
		Mesoproterozoic		1,600
		Paleoproterozoic		2,500
	Archaean			4,500

Figure 1.1. Geologic timescale.

The high-energy radiation, heat, and pressure created when bolides hit kill everything at the point of impact. Everything in the splash zone of vaporized and molten rock and solid debris is also killed. Surface waters of lakes and seas become acidified. Smoke and dust clouds block the sun's rays and cause a drastic drop in temperature. Bolide impacts are also thought to trigger large-scale volcanic eruptions that may, in the long term, have greater climatic influence.

It's possible the Chicxulub comet, which landed in the Yucatan Peninsula around 65 million years ago, caused the Cretaceous extinction, or it might have been due to large volcanic eruptions that churned forth huge lava flows in India.[3] Whatever the cause, the extinction event at the end of the Cretaceous period was the second largest in Earth's history. At least 85 percent of all species disappeared, including the dinosaurs. Some scientists suggest that heat stress associated with bolide impacts caused new species to emerge. This may be true, but many of the groups used as typical examples, including mammals and flowering plants, existed before the catastrophe, although they were far less numerous, less diverse, and much less widespread. (For example, the mother of all mammals, affectionately named "Jurassic Mother," a little 15-gram shrewlike animal that clambered about in the trees in search of insects to eat, first appeared about 160 million years ago.[4]) These groups had been in a state of "dynamic stasis," which means their form and distribution was relatively stable and resistant to change.[5] They were liberated from this state when the dominant cycads, ferns, and reptiles were exterminated in the Cretaceous catastrophe. The number and range of marsupials (i.e., the "pouched" mammals like opossums, wombats, and kangaroos) expanded first, but another climate catastrophe at the end of the Eocene epoch, about 34 million years later, provided an opening for the "Jurassic Mother" and all placental animals (i.e., animals that carry their young inside the mother's womb for long gestation periods and bear live young, like dogs, cats, horses, and humans) to diversify.

French naturalist Baron Georges Cuvier (1769–1832) did not accept the idea of evolution, believing instead that all species were created at the same time. However, he recognized that catastrophes could result in species' subsequent extinction. According to Cuvier, all organisms were so perfectly adapted to their conditions of life that any destabilization of their environment was detrimental. This made evolution impossible. Thus, when a climate catastrophe occurred, organisms not suited to the devastated or changed environment could not change and went extinct. Those that did possess appropriate characteristics then moved into the vacated territory and flourished.

In many ways, Cuvier was right: Natural catastrophes did cause mass extinctions. They resulted in the demise of dominant species that were well-suited to the previously stable environment. They also opened up new environments for the survivors; however, those survivors needed a combination of luck, physiological and behavioral adaptability, and intelligence in order to adjust to the changed environment, diversify, and become the new dominant species.

When did humans appear on the scene, and what did that scene look like?

It took quite a while for humans to appear. Our early ancestors were some of those lucky, adaptable survivors, but it was not until between about three and four million years ago that the bipedal apes – hominins like *Paranthropus* (an early, now extinct, bipedal hominin) and *Australopithecus* (an early bipedal hominin believed to be related to humans) – appeared on the scene in Africa.[6]

Until the mid-1980s, anthropologists believed that one of those hominins, *Homo erectus*, was our direct ancestor. *Homo erectus* appeared in Africa about 1.9 million years ago and left Africa for Europe and Asia around 1 million years ago. Later, some anthropologists thought that *Homo erectus* split into two species: *Homo erectus* and *Homo ergaster*. According to this theory, *Homo erectus* went east to Asia and later became extinct, whereas *Homo ergaster* began life along the shores of Lake Turkana, Kenya, some 1.9 million years ago and continued on in Africa to become our direct ancestor. Although *Homo erectus* disappeared relatively early in the west, the species persisted in China until about 230,000 years ago and survived in Java until around 50,000 years ago (for a timeline showing key early *Homo* species events see Figure 1.2).

But, surprisingly, the first evidence of ancestors that combined the traits of both *Homo erectus* and *Homo ergaster* with modern human traits has been found not in Africa, but in Spain and possibly Italy. *Homo antecessor*, potentially the common ancestor of both Neanderthals and ourselves, appeared about 800,000 to 900,000 years ago in a Mediterranean forest environment among oak and cypress trees, mammals, and birds very similar to those alive today. Our cousins, the Neanderthals, separated from our own evolutionary lineage at least 500,000 years ago. This heavy-set, big-brained, human-looking species migrated into Europe and Asia long before *Homo sapiens*. They roamed Europe and western Asia, according to some, alone and unchallenged until humans arrived.

It was not until about 200,000 years ago that *Homo sapiens*, in the form we think of as human, appeared in Africa. About 120,000 years ago, when warm, wet conditions encouraged the African rainforests and mangrove swamps to expand and dry forests and savanna to shrink, the vegetation and fauna from northeast Africa extended eastward into the Levant, motivating *Homo sapiens* to move eastward out of Africa, following the vegetation and fauna into what is now Israel. Because *Homo sapiens* were simply following an extension of

Period	Event	Age in thousands of years
Quaternary	*Homo sapiens* reach New Zealand	0.8
	Homo sapiens reach Hawaii & Easter Island	1.1
	H. sapiens reach Pacific Islands, *New Caledonia & Samoa*	3
	Aboriginal peoples inhabit Canada's Arctic	6
	Short cold event	8.2
	Holocene warm interval & first agriculture begin, megafauna extinction in N. America	10
	Homo floresiensis disappears	17
	Younger Dryas cold spell	12.7–11.7
	H. sapiens appear in the Americas	>20–10
	Coldest time of last ice age	21
	Neanderthals disappear	24.5
	H. sapiens in NE Russia	32–18
	H. sapiens in Tasmania	35
	H. sapiens appear in Europe, New Guinea	~40
	H. sapiens appear in Australia	~45
	H. erectus disappears in Java, cultural renaissance in Europe	after 50
	Behaviourally modern *H. sapiens* appear in China	~67–30
	Neanderthals in W. Russia until	~73–36
	H. floresiensis appears on Flores Island	74–38
	H. sapiens migrate out of Africa	~119
	Beginning of last glacial cycle	135
	Modern humans appear in Africa	~200
	Last *Homo erectus* in China	230
	Neanderthals separate from *Homo* lineage; *H. erectus* in Japan	by 500
	H. erectus? on Flores Island	840
	Wooden javelins in Germany	900
	H. erectus dispersal out of Africa	Prior to 1000
	Early stone tools	2300
		2600

Figure 1.2. Quaternary timeline.
Source: Adapted from Hetherington, R., and Reid, R.G.B. (2010). *The Climate Connection: Climate Change and Modern Human Evolution. Cambridge*: Cambridge University Press, p. 6.

their African habitat, these early human explorers did not have to substantially change their behavior to adjust to their new home.

Between 60,000 and 30,000 years ago the world's climate became very changeable. Extensive ice sheets stretched across the northern hemisphere, and a polar desert likely covered much of central Eurasia. These barren northern landscapes probably discouraged modern humans from straying too far north. Instead, by about 45,000 years ago, when global average sea level had dropped 70 meters below what it is today, they migrated along the southeast Asian coast and into Australia. It was not until about 40,000 years ago that *Homo sapiens* moved into Europe, and soon after that the climate worsened, becoming cool, dry, and very unstable. Persistent ice sheets to the north and an expanding polar desert to the east confined *Homo sapiens* and Neanderthals to a shrinking habitat in Europe. Neanderthals had survived earlier glaciations and subsequent warm periods, but this one was different. *Homo sapiens* seemed to thrive in the highly variable conditions, but Neanderthals did not. By 28,000 years ago, the last Neanderthals disappeared from central Europe, although they remained in what is now Portugal until as late as 24,500 years ago, when the world was solidly in the depths of the last ice age. About 20,000 years ago, when ice sheets up to 2 kilometers thick extended across much of northern North America, *Homo sapiens* moved into the Americas.

Homo sapiens migrated out of Africa and around the world during the last glacial cycle, a time between 135,000 years ago and about 11,650 years ago, when the world went through a series of ice ages and warm intervals (see Figure 1.3). Temperatures in Antarctica during that period fluctuated between 4°C warmer (when the glaciers were in retreat) and 10°C colder than today (when the glaciers had spread to their widest extent). During the cold periods, massive continental ice sheets covered northern North America, Greenland, Europe, and northern Asia. Global average sea level fell by as much as 120 meters as huge quantities of fresh water were locked up in glaciers. The large underwater shelves that extend out from the edges of the continents emerged from the sea. These exposed continental shelves became available for early humans to inhabit and travel along as they migrated from place to place.[7] One of them, the Beringia Shelf, linked northeast Asia to North America when sea levels were lower. Some say this subcontinent provided a route for humans as they moved into the Americas during the last ice age.[8]

Yet despite the erratic climate and their lack of protection from cold, hunger, and wild animals, humans not only persisted but spread throughout the world. By the onset of the worst and most pervasive

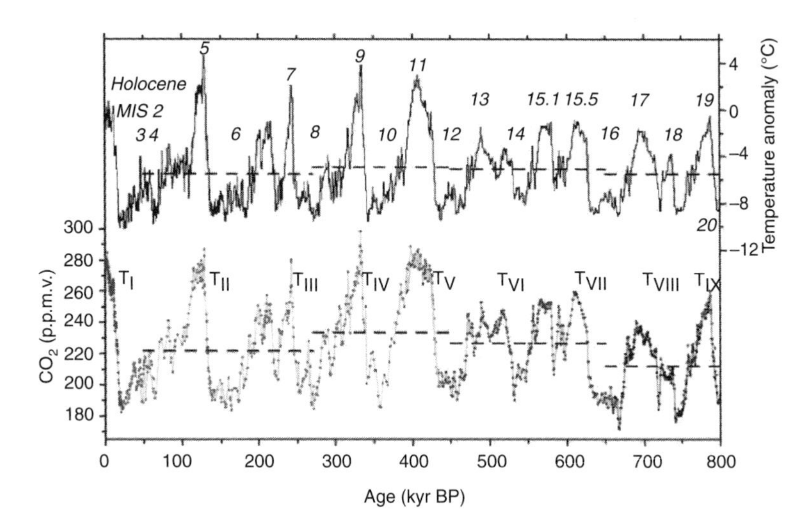

Figure 1.3. CO_2 records and temperature reconstruction of lower atmosphere over the last 800,000 years based on air trapped in ice cores from Antarctica. Temperature is expressed as an anomaly relative to present-day values. For example, during the depths of the last ice age, air temperatures over Antarctica were about 10°C colder than today. The data, described in Lüthi et al. (2008) and Petit et al. (1999), is derived from analysis of historical stable isotopes from the EPICA Dome C and Vostok ice cores.[9] Reprinted by permission from Macmillan Publishers Ltd. *Nature*

period of the last ice age (about 21,500 years ago) humans had colonized Asia and Australia, begun their migration to the Americas, and, as a result of the demise of the Neanderthals, become the sole *Homo* species in Europe. By 17,000 years ago, our last known *Homo* relative – *Homo floresiensis,* a three-foot-tall human that lived in Indonesia – went extinct.

Roughly 15,500 years ago, the climate began to warm rapidly as Earth cycled out of the last ice age. The enormous glaciers that soared kilometers in height over much of the northern hemisphere swiftly receded, leaving large glacial lakes in their wake. Global average sea level rose and the continental shelves sank beneath the sea. The homes and gathering sites of coastal humans living along the previously productive and relatively flat continental shelves were soon inundated by rising seas; people had to move inland, often to more hilly, mountainous, and forested terrain. Much of the marine food they had gathered and hunted was lost as their previous hunting and gathering grounds disappeared beneath the sea. As a result, food resources from the land likely became a more important part of their diet. An exception to this can be found along the Pacific northwest coast of Canada, where migrating salmon still swim up the rivers from the sea each fall, bringing rich marine resources that feed the forest, the wildlife, and the coastal people. These salmon rivers and spawning grounds became the territorial fishing grounds of early matriarchal coastal clans and were harvested much like an agricultural resource.

The warming 15,500 years ago was interrupted by a series of cold periods, the longest of which is known as the Younger Dryas. It lasted about 1,000 years and was caused by the rapid draining of a massive ice-dammed lake, Lake Agassiz, that covered much of central North America. Its fresh water poured into the Atlantic Ocean, disrupting currents and the circulation of the Atlantic Ocean. The Younger Dryas ended about 11,650 years ago. There have been cold events since then, but they have been much shorter and less drastic.

How did humans come to be the dominant species on Earth?

Humans became the dominant species on Earth after the Younger Dryas when we learned how to control and exploit plants, other animals, and nature generally. (Although some might dispute that bacteria are still more dominant.) An expanding human population discovered how to domesticate plants and animals; agriculturalists began to settle and produce food rather than remaining mobile to hunt and collect food. Independent centers of domestication developed about 10,500 years ago in southwest Asia, 500 years later in South America, 9,500 years ago in northern China, 7,000 or more years ago in Africa, and 5,500 or more years ago in Mesoamerica. Agriculture came much later to Europe and the eastern woodlands of North America.

Throughout these centers of domestication, agriculture gave rise to culture, tribal structure, and a sense of belonging to a particular tract of land, which in turn led to the concept of states or nations, complete with hierarchical and centralized decision making or government. Many people were no longer directly involved in collecting or producing food. Instead, people became full-time artists, architects, warriors, and politicians. Typically, economic systems evolved that required people to pay taxes and allowed those who did not produce food to purchase food from the agriculturalists. Early "scientists" and farmers developed new agricultural technologies, including crop rotation, the heavy plough, and irrigation. A political elite created labor and management systems to maintain resource production and irrigation systems. As the population grew, trade developed, along with competition and warfare.

These societal changes, combined with the conscious human drive to control and manipulate nature, ultimately resulted in the Industrial Revolution, which began just over 200 years ago. Humans developed technologies that made use of the concentrated energy captured in fossil fuels like coal, oil, and gas – fuels created by geological processes that concentrated the energy of earlier organisms after they died.[10] And we developed technologies that used increasing quantities of energy to overcome the hostile climate of previously marginal environments. With the exception of the oceans and the poles, we now inhabit virtually the entire planet.[11] Today, Earth's 7 billion people maintain our position of dominance and economic stability by using and controlling Earth's resources and the environment. We are increasingly concentrated in urban centers, and our attitude of dominance has widened the chasm between humans and nature, lulling us into a belief that we are protected from unexpected change and natural disasters.

Does dominance always prevail?

No. Just as previously dominant species, which flourished during periods of climatic stability, were wiped out by bolide impacts or volcanic eruptions and the resulting changes in climate, so too have human societies experienced periods in which they prospered and then declined. Humans have generally thrived in the relatively stable climate of the Holocene interval, which has lasted 11,650 years, but individual societies have disappeared. Hunter-gatherers were overridden and made marginal by the successful agriculturalists. The Maya civilization went through astounding expansion until its agriculture reached a critical threshold and climate change brought drought, at which point the cities fell and bedlam ensued. The Roman Empire conquered much of the known world between 31 BC and about 400 AD, then it grew weaker and contracted drastically. More recently, the British Empire expanded around the world, creating the largest empire ever known in human history, then it too suffered decline.

How is the climate change we are facing now different from past climate change?

In the past, climate change was caused by bolide impacts, volcanic eruptions, and changes in the amount of solar radiation coming into the atmosphere from the sun and/or the amount of energy emanating from Earth. These latter variations occur as a result of a number of important natural factors, including changes in Earth's orbit, variations in ocean circulation, and changes in the composition of Earth's atmosphere – for example, changing levels of greenhouse gases including carbon dioxide, methane, and water vapor.

The rapid climate change we are facing now is, to a vast degree, the result of humans pumping more greenhouse gases into the atmosphere than can be absorbed by Earth. This rise in greenhouse gases comes from our burning of fossil fuels and the changes we have made to Earth's landscape; for example, fewer forests exist to absorb carbon dioxide.

Typically, the land and oceans absorb as much carbon dioxide as they emit into the atmosphere each year. However, since the start of the Industrial Revolution, more carbon dioxide and methane have been emitted into the atmosphere each year than can be absorbed by the land and oceans. Today, seven more gigatons of carbon dioxide are added to the atmosphere each year than are absorbed by the land and oceans. Most scientists agree that this increase is caused by humans burning fossil fuels (see Figure 1.4) – including the burning of coal – as well as cement production and changing land-use patterns resulting from deforestation and agricultural processes. Given the present rate of economic and technological expansion, scientists believe that by 2100, humans will be sending 12 unabsorbed gigatons of carbon dioxide into the atmosphere each year.

Climatologically, the increase in emissions has resulted in a startling phenomenon – the atmospheric concentration of greenhouse gases is increasing at a drastic rate. Ice cores taken from glaciers in Antarctica contain minute air bubbles that have been locked in the ice sheet since its inception. Scientists have used these ice cores to determine the atmospheric concentration of carbon dioxide over the past 800,000 years. Figure 1.3 shows how these concentrations have oscillated over the past 800,000 years. Figure 1.5 shows how changes in CO_2 levels over the past 650,000 years compare with those of today and projected levels that scientists believe will be reached by 2100.

Scientists have also discovered that surface-air temperatures move in lockstep with the concentration of carbon dioxide and other gases like methane (see Figure 1.3).[12] Because of their greenhouse effect, carbon dioxide and methane levels make natural increases in temperature worse. Figure 1.5 shows that atmospheric carbon dioxide levels increased from about 180 ppm during the last ice age (21,500 years ago) to about 280 ppm in 1800 – when humans began increasing

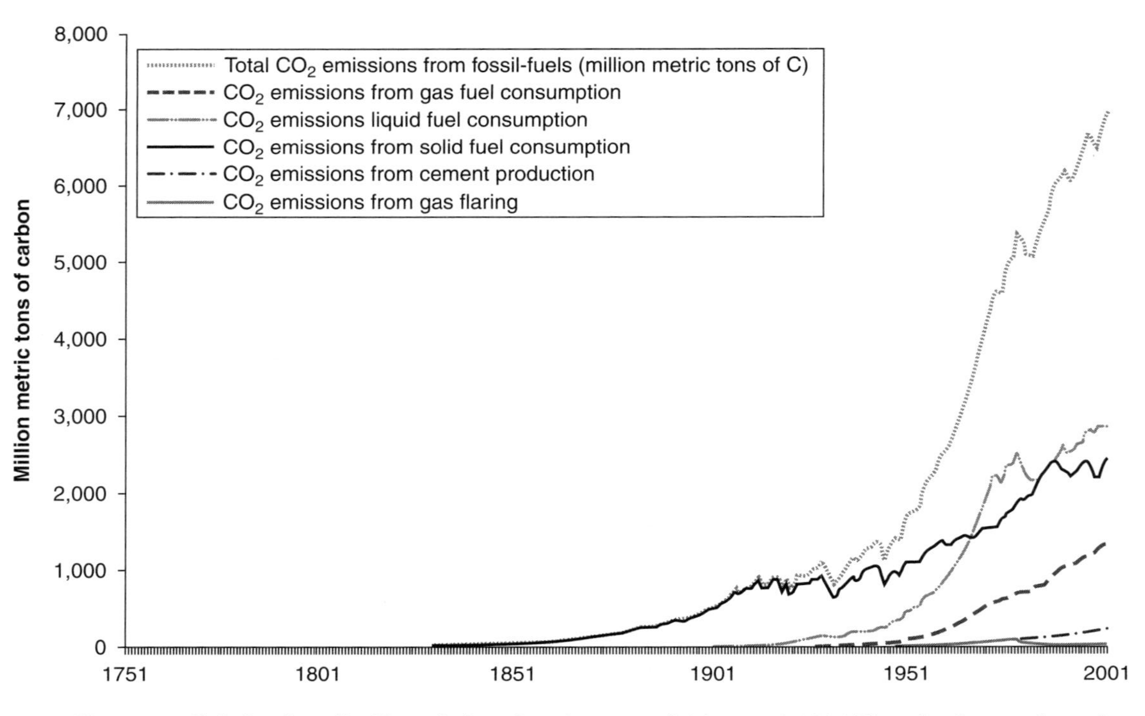

Figure 1.4. Global carbon dioxide emissions from human activities, 1750–2004. There has been a sharp rise in atmospheric carbon dioxide concentrations since the Industrial Revolution. Adapted from Marland et al., (2005).[13]

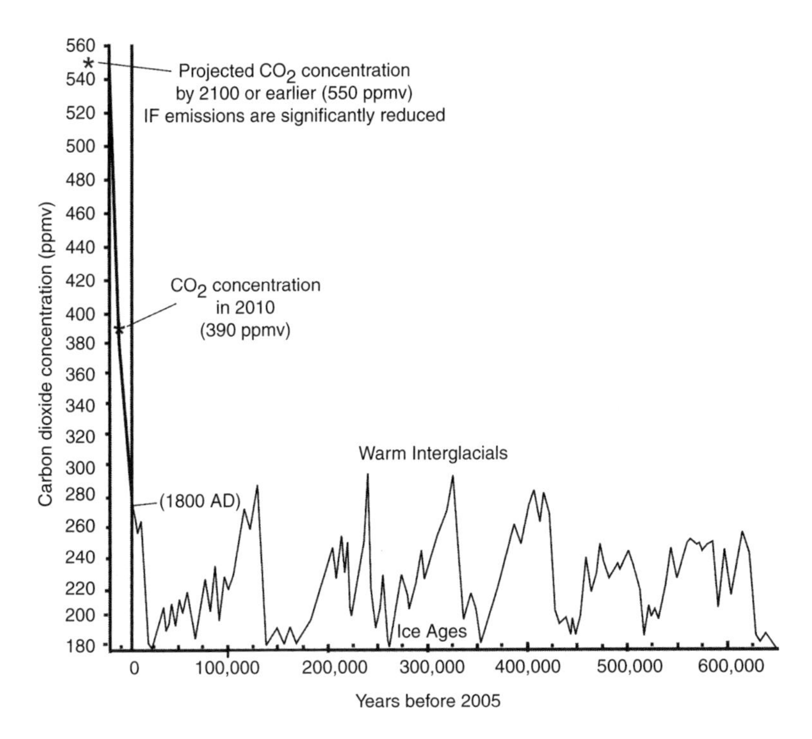

Figure 1.5. A composite atmospheric CO_2 record over 650,000 years – 6.5 ice-age cycles – based on a combination of CO_2 data from three Antarctic ice cores: Dome C, Vostok, and Taylor Dome. Adapted from Siegenthaler et al. (2005, figure 1.04)[14] and compared with atmospheric CO_2 levels in 2010 and levels projected to be reached by 2100 or earlier.

the amount of carbon dioxide in the air by burning copious amounts of fossil fuels – to 390 ppm in 2010. This means that carbon dioxide concentrations have increased by the same amount in the last 160 years as they did during the previous 21,000 years.

In a climate state that has not been influenced by humans, a 100 ppm change in carbon dioxide is the difference between an ice-age climate, with ice sheets covering virtually all of Canada and much of Europe, and what we consider to be Earth's present "normal" climate. It took 21,000 years for Earth's climate to naturally generate and adjust to a change of 100 ppm in atmospheric carbon dioxide. It has taken 160 years of burning fossil fuels to generate a similar increase, and the climate system is just beginning to adjust to this new load of carbon dioxide now in the atmosphere. Because the ice-core records do not show evidence of CO_2 levels reaching anything near 390 ppm over the last 800,000 years, many scientists believe that global average surface-air temperatures will rise to levels warmer than anything Earth has experienced over the last 800,000 years – warmer than anything that humans have ever experienced.

Recently, scientists discovered that many plants and animals are shifting where they live in response to climate change.[15] On average, they are moving 11 meters uphill and 17 kilometers toward the poles every 10 years, although some are moving much faster than others. As average temperatures continue to rise, the agricultural belt will likewise shift upward and toward the north and south poles. Species dependent on cold habitats will find themselves restricted to smaller and smaller areas. Increased aridity in the highly populated subtropical regions (which include the southern Mediterranean, northern Sahara, Australia, Florida and southern California, Brazil, and southeast China) will reduce the availability of drinkable water sources. People will be motivated to migrate out of drought-ridden regions in their search for food and water. During our early history, humans were able to migrate to new locations around the world in response to rapid climate change; today and in the future, however, political boundaries will restrict movement of an expanding human population, creating rising tensions across otherwise invisible boundaries.

Such change is difficult to imagine.

Why do we persist in ignoring the repercussions of our actions?

Although the number of extreme events and the variability of climate are escalating, the Western world – whose developed countries are, to an extraordinary extent, most responsible for escalating greenhouse gas emissions to date – has, thus far, experienced the least impact of climate change. This has undoubtedly allowed us to remain resistant to change.

"The weather is not so bad. Actually, it is pretty nice here where I live," climate change dissenters claim. In some ways they are correct. Weather is not climate. Weather is what you see out your window today; climate is the probability of that weather continuing to occur.

Other dissenters ask why, if carbon dioxide concentrations have increased so much in the past 150 years, the climate is still so similar to what it was during the Industrial Revolution. The reasons dwell in the oceans, which are the greatest "sink" for atmospheric carbon dioxide. The rate at which oceans absorb atmospheric carbon dioxide is much slower than the rate at which it is now accumulating in the atmosphere. Fifty percent of atmospheric carbon dioxide is absorbed in the top layer of the ocean within thirty years, although it often escapes back into the atmosphere. Another 30 percent of the carbon dioxide gets mixed into the deep ocean over centuries. (However, the more it absorbs, the more saturated it gets, and the less able it is to absorb more.) The last 20 percent of the carbon dioxide stays in the atmosphere for thousands of years.

As mentioned earlier, carbon dioxide in the atmosphere creates a greenhouse effect that makes natural increases in temperature worse. The more carbon dioxide that remains in the atmosphere, the warmer the climate gets. And this warming happens quite quickly.

Also, as the oceans absorb carbon dioxide, their chemistry and currents change. This alters the climate too, but it happens over the longer term.

So even if we do not feel climate change yet, we know it is coming.

What role does evolution play in our ability to genetically adapt or adjust to climate change?

When discussing evolution today, it is important to contrast two theories of evolution. The first, Darwinian theory emphasizes gradual genetic change through natural selection over geological time. Darwin's work treated adaptation to climate change as part of evolution, but he did not delve deeply into physiological and behavioral adaptability, factors that are crucial if individuals are to survive severe environmental changes.

It is important here to make the distinction between "adaptation" and "adaptability." Adaptations are random, genetically fixed mutations that occur through Darwinian natural selection. For example, developing a shorter, squatter, and hairier body is an adaptation to the cold. Adaptability, on the other hand, is a physiological or behavioral modification by the individual. An example of adaptability would be putting on warm clothes and lighting a fire when the temperature drops.

As a gradualist, Darwin paid little attention to catastrophic events as he developed his theory; he did not examine the effects that the environment has on evolution. And although Darwin assumed humans were affected by climate, he did not explore the ways that humans can change climate.

The second theory of evolution is Emergence theory, which treats evolutionary change as a saltatory (or rapid) process. According to this premise, the organism changes in response to variations in the environment. Emergence theory describes the history of life as long stretches of time when communities and ecosystems are stable and nothing much - evolutionarily speaking - happens, punctuated by periods of rapid evolution. During the long phases of equilibrium, "natural selection" operates. During periods of stress and climatic or environmental instability, major emergences or changes occur.

According to Darwinian theory, evolution can occur in dominant species, but genetic adaptations occur very gradually over long periods of time. The theory also suggests that natural selection is not evolutionarily creative, and when species become highly adapted to prevailing conditions, novel change is resisted. This means that natural selection actually limits the ability of dominant, highly adapted species to change when environmental conditions change rapidly, because there is not enough time for the species to genetically evolve. If the environmental change is significant enough, the species can go extinct. According to this view, evolution is unlikely to save the dominant *Homo sapiens* from impending rapid climate change.

On the other hand, Emergence theory recognizes that crisis or catastrophe may actually stimulate rapid genetic, physiological, and behavioral changes. By interacting with its new environment, the organism adjusts and generates new, more appropriate qualities. When applied to humans, Emergence theory suggests evolution is not only biological, but also cultural and social. Throughout human history, we have seen new behaviors emerge in the areas of cognition, language, tool manufacture and technology, socialization, aesthetic and religious awareness, education, and agriculture. Rapid changes in these areas are followed by periods of dynamic equilibrium in which evolutionary changes are rare or are resisted by prevailing traditions. In this scenario, climatic instability and social crisis may turn out to be opportunities to experiment with new ways of doing things. When defined according to Emergence theory, an evolutionary response to future climate change sounds a little more promising for humanity.

What does that mean for us today?

Today, our desperate attempts to resist environmental instability, and associated economic instability, by controlling our environment through voracious use of fossil fuels and unlimited consumption are feeding an increasingly rapid climate change while simultaneously nourishing our deluded confidence that, unlike all prior dominant species on Earth, our survival is guaranteed. We continue to rely on the Darwinian belief in "survival of the fittest," but as was the case with the previously dominant dinosaurs, survival of the fittest will not rescue humans from rapid climate change. If we depend on natural selection to drive our adaptation to the new environment, we are not likely to survive. Slow, gradual evolution will not operate quickly enough to allow us to adapt.

So what will?

When conditions change rapidly, innovative behavior and ideas combined with a population willing and able to accept and implement change improve survival.[16] A study of human history shows that crisis, communication, and collaboration – the three C's – act as an impetus for human social evolution, and social evolution is what might allow us to adjust successfully to future climate change. However, we will only change if we accept that we have a responsibility and the capacity to facilitate and make that change happen.

Emergence theory suggests that interaction with the environment actually helps create change. In the past, species that survived a rapid environmental change were those that experienced rapid or saltatory evolution. According to Emergence theory, then, rapid climate change also acts as an impetus for species change.

A caveat

Let there be no mistake: Earth's climate has always been changing. Over the last 800,000 years, Earth has experienced ten glacial cycles alternating between cold intervals – ice ages – and warm intervals – interglacials. During the warmest intervals, average global temperatures were up to 4°C warmer than today. The cold intervals were as much as 10°C colder than today. The last warm interval began about 135,000 years ago, not too long after the appearance of the first *Homo sapiens*.

Throughout the history of Earth, catastrophic climate change has also been a reality. As a consequence of these sudden changes, various species that once dominated became extinct, allowing previously marginal species to develop and dominate.

We are the last remaining *Homo* species on Earth, and we have achieved a dominance over Earth and its environment that is unprecedented. As a result of our activities, global biodiversity is plummeting;

the cultures and languages of nondominant human societies are also fast disappearing. The practices of dominant human societies are unsustainable, especially given the rapidly expanding global population. Although Earth's climate has continuously changed in the past, we are now influencing and exacerbating that change. Atmospheric levels of carbon dioxide will soon be higher than they have been since the Mesozoic era – the Age of the Reptiles.

Like all other species that have lived on Earth, we have the capacity to adjust our behavior, a capacity limited only by our intelligence, and this remains our last vestige of hope. We need to change our behavior in order to generate an immediate large reduction in greenhouse gas emissions and to restrict our consumption to that which Earth can sustainably provide. But the desire of people in developed countries to maintain gluttonous consumer habits, and the desire of people in less economically wealthy countries to obtain the goods, services, and economic prosperity of wealthier societies, suggests it will be difficult to alter our behavioral repertoire, our technological tool kit, and our adherence to current paradigms.

Before we delve more deeply into the changes that are coming and the changes we must make, let us explore more thoroughly the history of our ancestors.

The Evolution of the *Homo* Species

2

The Cradle of Humankind

Endless forms
most beautiful and most wonderful have been,
and are being evolved.

Charles Darwin (1809–82), On the Origin of Species

One day, as the story of the first human emergence might go, an early human ancestor stood upright amid the grasslands of the African savanna and staggered a few steps before tumbling among her astonished playmates. A game ensued as they all began to balance on their hind legs and poke their tiny apelike heads above the tall grass. At one point when the original curious one began to fall, a gallant young male reached out and took her hand. Soon the two were leaning against each other, chuckling and tottering away.

It was not long after that a group of our early ancestors came upon a recently killed carcass. They were chattering away in their guttural language as they ripped open the hide and stuffed meat into their mouths. They pushed each other aside, vying for a chance to tear meat off the bones; a food fight ensued. Meat, blood, and bones flew through the air. One rambunctious adolescent, who had just been whacked in the head with a bone, turned and heaved the closest thing at hand, a cobble. It narrowly missed the aggressor, crashed into another rock, shattered, and ricocheted through the air. A large splinter of rock lodged in the aggressor's leg. The food fight abruptly halted. They looked on in amazement as blood began to pour out of the large gash. Bending over, the adolescent pulled out the piece of rock, but instead of throwing it away as one might expect, he looked from the bloodied rock in his hand to the carcass and back. Bending over the carcass, he ran the sharp edge of the broken rock along the hide. To the astonishment of his friends, a chunk of meat slipped neatly to the ground. Silence, and then another frenzy of excitement ensued. Rocks began flying in all directions as each tried to make their own broken rock chunks to hive meat from the rapidly depleting carcass.

Although this is just a story, the actual circumstances that led to our learning how to walk and create tools may be similar. We will never know with absolute certainty the origins of these skills, but it is clear that early humans learned how to walk and possessed a spark of creativity that helped them develop an increasingly complex collection of tools and new technologies. These characteristics helped us survive and flourish in a frequently harsh and changing environment.

Where did that spark of creativity begin? Where did that cradle of humankind rest, and who was in it?

Aristotle placed humans at the top of the Ladder of Life, above apes and other animals. It was because we were the most complex organism that he ranked us on the top rung. The simplest plants were positioned on the bottom. But Aristotle did not recognize how one type of organism could evolve or change into another. It was George Louis Leclerc Buffon (1707–88) who stated: "If, for example, it could be once shown that the ass was but a degeneration from the horse – then there is no further limit to be set to the power of nature, and we should not be wrong in supposing that with sufficient time she could have evolved all other organized forms from one primordial type."[1]

In 1871, Charles Darwin declared that Africa was the cradle of humankind.[2] He was likely spurred on by his supporter, biologist Thomas Henry Huxley, who interpreted Darwin's theory of evolution to mean that humans had evolved from apes.[3] Darwin believed that a series of progressively more complex species evolved in a linear progression from apes to modern humans. At some time along this progressive evolutionary ladder, the defining moment that marked the beginning of the history of modern humans occurred – the appearance of stone tools. According to anthropologist Richard Potts, Darwin built his scenario of human emergence on our tool making ability, our social abilities, and our habitation of the ground, as opposed to trees.[4] Life on the treeless savanna meant early humans were more vulnerable. We lacked the large eyeteeth of our predators, so we needed to create tools to protect ourselves. With tools, early humans were able to survive on the grassy savanna, where the ability to walk upright helped them see the prey they were hunting and the predators they were trying to avoid. Following these developments – perhaps stimulated by them – humans developed an enlarged brain and, later, culture and social contracts.

For over a century after Darwin published his theories, archaeologists thought that the fossil record supported his belief that humans evolved along a linear, staged progression. A series of so-called transitional forms represented the stages of progression from ape to human. Each new form was believed to display a new physical (morphological) transition, each building on the other and all in all spanning the long evolutionary distance between apes and humans. This linear progression meant that there existed no *Homo* species diversity. Instead, humans evolved from apes, no other species evolved along this path except humans, and only one *Homo* species existed at any one time.

Today, biologists commonly recognize that many species possess a great deal of species' diversity (think of all the different species of fish that have been identified), but this unilinear progressionistic perspective dominated the idea of how humans evolved until the 1960s. It was then that archaeologists uncovered evidence in Africa of more than one, and potentially three, contemporaneous hominid species. (Note: The term *hominid* here refers to all the great apes, whereas *hominin* refers to the bipedal apes, including all the fossil species and living humans.)

The archaeologists found that about 6 million years ago the hominid line diverged into two branches. One branch split into chimpanzees and gorillas. From the second branch, *Australopithecus* species (meaning "southern ape") emerged about 4 million years ago or earlier. Many archaeologists and anthropologists believe that our ancestor is *Australopithecus afarensis*. This apelike species lived in Africa at a time when grasslands in eastern and southern Africa were becoming more extensive and forested areas were shrinking. They base this theory on fossils discovered at various locations in Africa (see Figure 2.1). So according to fossil evidence we now know that humans are not descended from apes – that is, chimps and gorillas. However, we do share a common ancestor.

About 2.4 million years ago, in the Ethiopian River Valley in Africa, a small human, apelike creature emerged from *Australopithecus*, probably from *Australopithecus afarensis*. Archaeologists called this new hominin *Homo habilis*, or "handy man," because it was the first hominin known to use complex stone tools, although stone tools have now been found that are slightly older than the oldest evidence of *Homo*. This new species also had a slightly larger brain than *Australopithecus afarensis*. Our ape cousins may have looked on one day in puzzlement

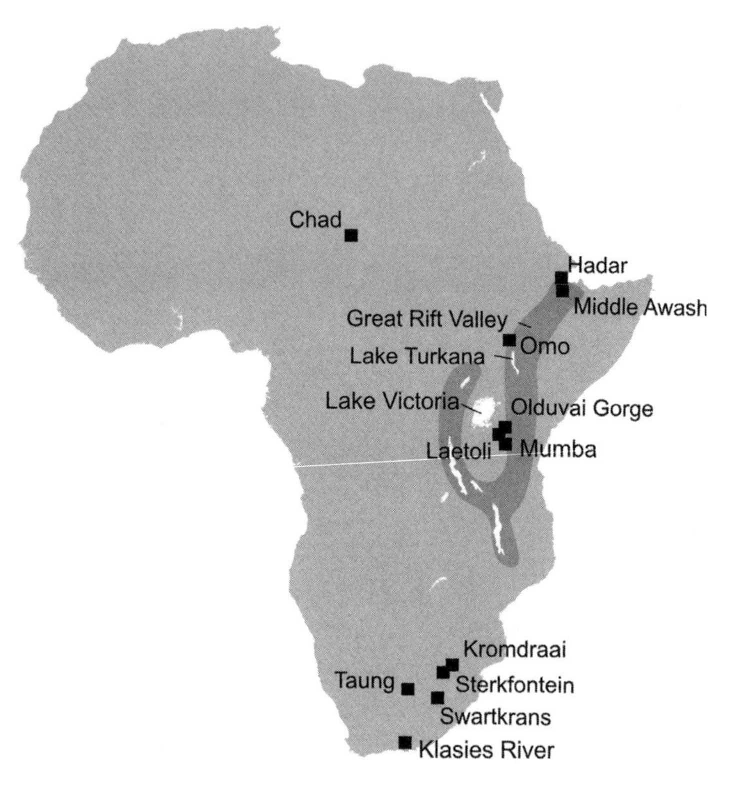

Figure 2.1. Key early hominin sites in Africa.

as *H. habilis* picked up a stone, used it as a hammer, and then cut meat from bones. Archaeologists call these first stone tools the Oldowan technology, after the Olduvai Gorge in Tanzania where they were first discovered. Unlike the earlier apelike australopithecines that lived in mixed grasslands and woodland environments and survived mainly on a vegetarian diet, *H. habilis* ate an omnivorous diet and lived in grasslands beside lakes and rivers.[5] Meat was becoming an increasingly important part of *H. habilis*'s diet. It is clear to us now that *H. habilis* parents taught their children how to make stone tools; this knowledge and the skills that went with it were passed down from generation to generation, remaining unchanged for a million years.

Many fossils have been unearthed, including those of *Australopithecus afarensis*, our potential early ancestor, along with other early *Australopithecus* and other similarly ancient species that are not our ancestors. Table 2.1 lists many of these as well as early *Homo* species, most of which we could call our cousin species, but not our direct ancestors.

Table 2.1. *Hominin species*

Australopithecus afarensis lived 3.85 million to 2.95 million years ago in eastern Africa. Its best-known representative is the skeleton named Lucy, who was 1 meter tall, weighed about 27 kilograms, had a brain less than one-third the size of that in modern humans (380–450 cc), and was found by Donald Johanson and his team of paleoanthropologists in 1974 at Hadar, Ethiopia. This site is also known for the oldest documented bipedal footprints - of adults and a child in now-hardened volcanic ash - discovered by Mary Leakey and paleoanthropologist Tim White at Laetoli in northern Tanzania in 1978. *Australopithecus afarensis* had long arms and curved fingers that helped them climb trees, although they also walked upright on the ground. These adaptations helped them survive almost a million years as the climate and environment changed. Many archaeologists believe *Australopithecus afarensis* is the ancestor of the *Homo* species.

Australopithecus africanus lived 3.3 million to 2.1 million years ago in southern Africa. With their ape- and humanlike features, they were very similar to *Australopithecus afarensis*, although slightly taller and heavier. They had a brain size between 430 and 550 cc, walked upright, climbed trees, and ate plants. They were first discovered by Raymond Dart in 1924, when he uncovered the skull of a three- to four-year-old child in South Africa. Archaeologists also consider *Australopithecus africanus* a good candidate to be the ancestor of the *Homo* species.

Australopithecus anamensis lived 4.2 million to 3.9 million years ago in Lake Turkana, Kenya (*anam* means *lake* in the Turkana language). Bones of *Australopithecus anamensis* have also been found in a wooded environment in the Awash Valley of Ethiopia. A bipedal hominin, believed to be transitional between apes and later australopithecines (according to Tim White, who asserts a single evolutionary lineage of humans that began in East Africa, *Australopithecus anamensis* is a direct descendent of *Ardipithecus ramidus*), they ate plants and nuts and lived in forests and woodlands around lakes.

Australopithecus robustus lived 1.8 million to 1.2 million years ago in South Africa. They were originally found and named *Paranthropus robustus* by Robert Broom, a Scottish medical doctor and amateur paleontologist, in the Kromdraai Cave, South Africa, in 1938. Males grew to 1.2 meters tall and averaged about 54 kilograms; women were about 1 meter tall and weighed 40 kilograms. *Australopithecus robustus* had a sagittal crest - a ridge of bone that ran along the top of the head from front to back - to anchor the big chewing muscles they used to crush and grind tough food like nuts, seeds, and roots, although they also consumed fruit and possibly meat and plants. Another robust *Australopithecus* is *Australopithecus aethiopicus*, found by Alan Walker on the west side of Lake Turkana in 1985 and dated to 2.5 million years ago.

(*continued*)

Table 2.1. *(continued)*

Australopithecus sediba lived 1.977 million to 1.78 million years ago. The fossilized skeleton of a child and several bones of an adult of this most recent *Australopithecus* were discovered by Lee Berger and others in the Malapa caves near Sterkontein in what is now South Africa. The young male was 1.3 meters tall and had a brain size similar to a chimpanzee. *Australopithecus sediba*'s long arms would have been good for climbing, but their hips were similar to those of modern humans, so they probably walked on two legs. They may have eaten more meat than earlier *Australopithecus* or used stone tools to process plants to make them easier to chew. Berger suggests *Australopithecus sediba* may be descended from *Australopithecus africanus.*

Homo erectus, "upright human," lived 1.89 million to 143,000 years ago, and as late as 50,000 years ago in Java, about nine times longer than *Homo sapiens* have been on Earth. They lived in Africa, western Asia, China, and Indonesia; stood between 1.45 meters and 1.85 meters tall; weighed between 40 kilograms and 68 kilograms; and had a brain size of between 700 and 1,250 cc. They were first discovered and named *Pithecanthropus erectus* by Eugène Dubois in 1891 in Indonesia. The most complete skeleton is Turkana Boy from Lake Turkana, Africa, dated to 1.6 million years ago. *Homo erectus* is believed to be descended from *Homo habilis.* They lived on meat, other protein, and possibly honey and underground tubers. They used the Acheulean stone tool industry, including hand axes and cleavers that helped them adjust to a changing climate and become the first *Homo* species to expand out of Africa. They used fire for cooking, staying warm, and keeping predators at bay.

Homo ergaster lived 1.9 million to 1.5 million years ago. Some consider this a separate species from *Homo erectus.* It originated at Lake Turkana, Kenya, 1.9 million years ago and remained in Africa. Some scientists believe they are the direct ancestor of *Homo sapiens.* Their brain size was up to 850 cc.

Homo floresiensis, "The Hobbit," lived 95,000 to 17,000 years ago in Indonesia. They stood 1.06 meters tall, weighed 30 kilograms, and had a brain size of 380 cc; they hunted small elephants and large rodents, lived in proximity of the giant Komodo dragons, and may have used fire. *Homo floresiensis* is thought to have evolved from *Homo erectus* (which may have been the hominin species that made the 800,000-year-old stone tools also found on the island of Flores) or, possibly, another small early species of the genus *Homo.* They or their predecessors had to make a dangerous sea crossing to reach Flores.

Homo habilis, "Handy man," lived 2.4 million to 1.4 million years ago in eastern and southern Africa. Around 1 meter to 1.35 meters tall, they weighed, on average, 32 kilograms and had a brain size of 600 to 800 cc. They were first discovered by Louis and Mary Leakey between 1960 and 1963 at Olduvai Gorge in Tanzania, along with thousands of stone tools called the Oldowan technology that date to between about 2.5 million and 1.2 million years ago. Its face does not protrude as much as earlier hominins, but it is still very apelike. *Homo habilis* lived on an omnivorous diet in grasslands by lakes and rivers in Africa.

Homo heidelbergensis lived 700,000 to 200,000 years ago in Europe, Africa, and possibly China. Males averaged 175 centimeters tall and weighed 62 kilograms; females averaged 157 centimeters and weighed 51 kilograms. They had a larger brain size than modern humans, controlled and used fire, and were the first early human species to hunt large animals and build dwellings out of rock and wood.

Homo neanderthalensis separated from our lineage about 500,000 years ago and lived from at least 200,000 to 24,500 years ago in Europe and southwestern and central Asia. Males averaged 164 centimeters and females averaged 155 centimeters tall. They were shorter and stockier than modern humans, with brow ridges and a larger brain (1,450 cc). They hunted big game with wooden spears, then butchered the animals to eat and sewed the hides for clothing. They also ate plants. Neanderthals who lived along coastal regions ate mollusks, seals, fish, and other marine resources. They contributed the Mousterian stone tool technology, which unlike earlier "core tool" technologies involved detaching flakes from a prepared stone core and using those flakes to create new stone tools. They used fire, built shelters, and practiced symbolic behavior, including burying their dead with offerings.

Homo sapiens, "Wise Man," first appeared about 200,000 years ago. They evolved in Africa before moving out of Africa about 120,000 years ago and migrating around the world. They have a slightly larger brain than *Homo erectus* (1,300–1,400 cc) and a rounder skull, smaller teeth and jaw, high forehead, prominent chin, and no brow ridges. The term "anatomically modern *Homo sapiens*" or "anatomically modern humans" (AMH) refers to members of the *Homo sapiens* species who lived during prehistoric times. These AMH made and used specialized stone and bone tools including bows and arrows, fishhooks and harpoons, spear throwers and sewing needles to hunt animals, gather plants, and process them to feed and clothe themselves. Just like their modern counterparts, they controlled fire, cooked food, made shelters, created art and music, and practiced symbolism.

Sources: Smithsonian National Museum of Natural History Web site, "What Does It Mean To Be Human?" at http://humanorigins.si.edu/evidence/human-fossils/species; "Discovery of Early Hominims" on Dennis O'Neill's Early Hominim Evolution Web site at http://anthro.palomar.edu/hominid/australo_1.htm; and a video by Stephen Churchill from Duke University on *Australopithecus sediba* at http://www.youtube.com/watch?v=pJOOo9C0dYE

Despite the fossil evidence of more than one early great ape and numerous *Australopithecus* species, scientists generally believed until the mid-1980s that a single upright walking hominin species, *Homo erectus,* left Africa for western and eastern Asia around 1 million years ago. Then a second theory emerged, suggesting that *H. erectus* had split into two species: *H. erectus* and *Homo ergaster.* According to this theory, *H. erectus* went east to Asia and later became extinct.

But how do we know which theory is correct?

Most archaeologists and anthropologists create their ideas of what modern humans were like and how they behaved from scientific interpretations of early stone tools that may or may not have been associated with fossil bones. They then develop theories from the tool and fossil evidence they have at their disposal. To gain a better idea of the kind of life one of our early ancestors led, let us spend a little time in the archaeologists' world of fossil bones and tools during the time of *H. erectus.*

Nearly 2 million years ago, the new, larger-brained *H. erectus* species evolved. Although the first *H. erectus* individuals are thought to have emerged in Africa, the earliest fossil evidence has been uncovered in Asia, suggesting *H. erectus* was an adventurer and traveler. Pakistan became the home of *H. erectus* perhaps as early as 2 million years ago.[6] *H. erectus* also lived in Georgia, in westernmost Asia, between 1.7 and 1.9 million years ago,[7] and in Java, southeast Asia, at least 1.5 million years ago.[8] *H. erectus* had a larger brain and was more innovative than *H. habilis,* and by 1.5 million years ago these pioneering hominins had created a new tool kit called the Acheulean, which contained large cutting tools, including teardrop-shaped, two-sided hand axes and cleavers. The hand axes may have actually been more like discuses. Early hunters threw these large stone projectiles fifty meters, striking the back or shoulder muscles of large prey animals, taking them down as they ran in herds. The Acheulean tool kit was made by *H. erectus* and also by *Homo heidelbergensis*, another larger-brained species in Europe. For reasons unknown to us, Acheulean tools stopped being made in northern Europe 500,000 years ago, but it was another 200,000 years before they disappeared elsewhere.[9] Around the same time Acheulean tools were developed, an ingenious *H. erectus* in Swartkrans Cave in Africa generated a spark, starting the first *Homo* species-made fire, the remains of which archaeologists have discovered.

In China, archaeologists have found stone tools that are 1.36 million years old and very similar to the Acheulean tool kit except that the teardrop-shaped hand axe is missing. Some say that the early *H. erectus* living in China made another tool for the same purposes out of something other than stone, perhaps bamboo, but it has not been preserved in the archaeological record. We may never know, but it is clear that 1.36 million years ago these early inhabitants of China were using tools to deal with very challenging seasonal environments.[10] A little more than 500,000 years later, early *Homo* species living in the Bose basin, which lies between the Loess Plateau and the South China Sea, witnessed what must have been a terrifying sight. A large meteorite struck the land. Fire engulfed and destroyed large tracts of forest. A crisis was at hand, and the choices they faced were stark – adjust their behavior to survive the harsh new living conditions or die. We now know that *H. erectus* and other early *Homo* species survived by changing their behavior and developing new stone tools.[11]

About 400,000 years after the meteor struck China, a troop of *Homo heidelbergensis*, or one of their close relatives, were living in a wooded site by a lake in what is now Schöningen, Germany. They cut down a number of thirty-year-old spruce trees (*Picea* sp.) and then carefully carved spears out of the trunks. The tips of the spears were made from the base of the tree, where the wood was hardest. To make a good spear, the distribution of weight is very important. These hunters made their spears with the same proportions as those of modern javelins – the center of gravity falling one-third of the way down the shaft. Then they planned their hunt; they took their beautifully carved spears, waited for their prey, then struck, killing and subsequently butchering more than ten horses.

Based on this and other archaeological evidence, it is clear that early humans responded to changes in their environment by adjusting their behavior. They made tools out of stones and trees or shrubs to help them kill animals for food. They migrated to new habitats in search of new territories to exploit and in response to changing conditions in their home territory. They made fires to keep warm, and, despite our previous misconceptions about the limited abilities of early hominins, some were clearly able to make advanced tools and plan hunting and subsistence strategies. However, in the end, *H. erectus* was not able to continue to adjust and went extinct around 143,000 years ago or later. No one knows why, but it was probably because they did not have the intellectual capacity or creative spark that allowed them to change their behavior as much as they needed to.

H. ergaster, the other *Homo* species that is considered a cousin species of *H. erectus*, lived along the shores of Lake Turkana, Kenya 1.9 million years ago.[12] According to the second theory, which had *H. erectus* splitting into two species, *H. erectus* and *H. ergaster*, the latter survived to become our ancestor. But this theory also has its problems. At two sites, one in the Sierra de Atapuerca of north central Spain and the other in southern Italy, fossils from early hominins have been unearthed that have modern facial characteristics and the primitive dentition of *H. ergaster* and *H. erectus*. Archaeologists say these fossils, which are more than 780,000 years old in Spain and between 800,000 and 900,000 years old in Italy, belong to a new species, *Homo antecessor*.

If this is correct, then a new species of *Homo* emerged around 900,000 years ago in Africa or Eurasia. This was the same time that a major early *Homo* dispersal event occurred out of Africa. To further add to the complexity, million-year-old fossil skulls (crania) found in Middle Awash, Ethiopia have characteristics that overlap those of the African *H. ergaster* and the Asian *H. erectus* samples. These finds have led some archaeologists to argue that splitting *H. erectus* into separate species is misleading.[13]

Elsewhere, technological and cultural evolution accelerated. By at least 100,000 years ago, and perhaps as early as 300,000 years ago, the Middle Palaeolithic tool industry appeared in Europe along with the coincident sub-Saharan African tools of the Middle Stone Age (see Figure 2.2). The large cutting tools of the Acheulean were replaced by smaller tools made using a new technique called Levallois and radial core technology. Small blades were hafted onto wooden handles; stone tips were placed on spears; knives and scrapers were mounted on shafts and handles. Making these kinds of tools requires what is called sequential nonrepetitive fine motor control. Speech also requires non-repetitive fine motor skills. Scientists think that both these skills are controlled by adjacent parts of the brain.[14] This has led some researchers to suggest that this new tool industry evolved at the same time as grammatical language, around 300,000 years ago in the Middle Palaeolithic, and that the common ancestor of Neanderthals and our own species could speak.

And so the search for the origin of the first modern humans continues.

As a result of these discoveries and a great deal of research done by generations of archaeologists and anthropologists, the definition of what it was and is to be human was recast. The new criteria became absolute brain size (600 cc), perceived capacity for language ability (as inferred from casts of skulls), hand function (precision grip and opposable thumb), and stone tool making. However, as archaeologists and anthropologists dug deeper, these criteria became unsatisfactory for a variety of reasons. The biological significance of brain size is questionable. Research shows that when scientists group fossil hominin skeletons by brain size, they end up with different groups than when they group the same fossils by body size and shape and by how the early hominins walked. Another reason is that the function of language cannot be reliably inferred from brain appearance, nor are language-related parts of the brain clearly localized. Further, the modern humanlike hand grip cannot be restricted to *Homo,* and early

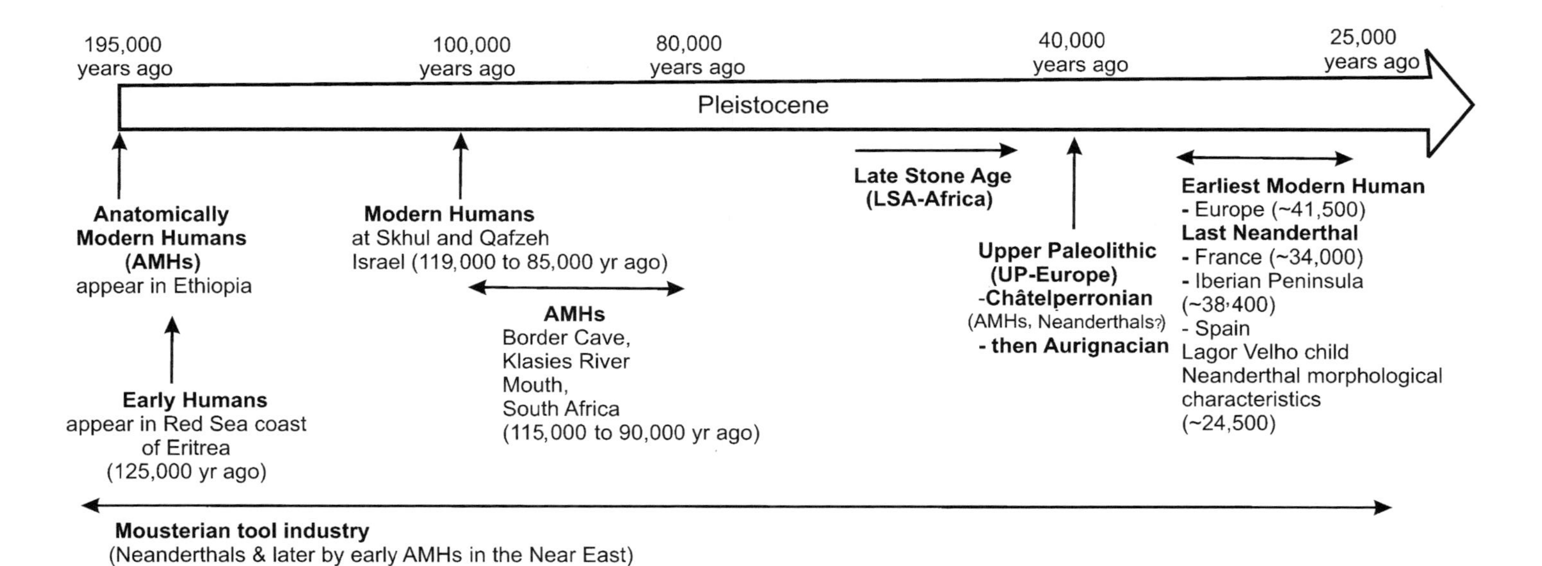

Figure 2.2. The fall of Neanderthals and the rise of modern humans.

Source: Adapted from Hetherington, R., and Reid, R.G.B. (2010). *The Climate Connection: Climate Change and Modern Human Evolution.* Cambridge: Cambridge University Press, p. 49.

stone tools dating to more than 2.3 million years ago are contemporaneous with both *Homo* and *Paranthropus* (an early, now extinct bipedal hominin).[15]

The fossil evidence is not clear. Increasingly, researchers are looking to genomists, people who study the genetic information of organisms, to explain the difference between humans and apes and other organisms through genetic variations. Yet, despite many recent scientific advances in genetics, little detailed knowledge of what sets early humans apart from apes has been revealed.[16] We also do not understand very well how developmental variations like language, longevity, modes of learning, or aging set us apart, and there are no genes for these characteristics.

We do know that there is only a 1.2 percent difference in the DNA at the nucleotide level between a bonobo chimpanzee (*Pan paniscus*) and a human.[17] This is not much more than the average 0.2 percent range of variation within *H. sapiens*,[18] and less than the variation among chimpanzees. The most distinct differences between, say, bonobo and human – hair loss, increased ability to walk upright on two legs, new hand anatomy, reduction in facial bone growth, and relative expansion of the neocortex[19] – all mainly involve differences in the distribution in space and time of hormones and their receptors. Those molecules do not need to change at all in their structure to produce the effects that result in the differences we see between us and the bonobo. This means that genetic explanations about what makes us human are limited.

If there is so little difference between ape and human genes, what does make us different? What do we know about humans – *Homo sapiens*?

The first anatomically modern *H. sapiens* appeared in Israel about 195,000 years ago and in southern Africa about 115,000 years ago. These humans looked much like we do today – without the chic haircuts and expensive clothes. But archaeologists believe it wasn't until about 70,000 years ago in Africa, and between 50,000 and 40,000 years ago in Europe, that our early human ancestors stopped behaving archaically and developed that spark of creativity that we identify with present-day humans.[20] During this time, the human brain did not get bigger or more complex in its overall anatomy. However, it could have reorganized at the cellular level. Some parts of the brain may have become differentiated, perhaps with an increase in the number of neurons and a more complex coordination. It is possible that the potential for that spark of creativity existed with the first *H. sapiens* that emerged, but it was only realized with the onset of suitable environmental, psychological, and social developments.[21] Once the right conditions occurred, a smart, modern human lineage could have rapidly increased its numbers. They may or may not have interbred with the original, less creative *H. sapiens* or with surviving *H. neanderthalensis*. But the result was that a new creative *H. sapiens* predominated and *H. neanderthalensis* died out, probably for a variety of reasons: a rapidly changing climate, an inability to adjust quickly enough, and a mean streak that came along with the newly creative *H. sapiens*.

We have uncovered an archaeological history that tells us of a series of ancestors and cousin *Homo* species that, for reasons we still don't fully understand, emerged, lived, and then, one by one, became extinct. Let us now look at how we have come to be all alone – the last remaining *Homo* species.

3

The Neanderthal Enigma

> I know not with what weapons World War III will be fought, but World War IV will be fought with sticks and stones.
>
> *Albert Einstein*

It is Europe, 30,000 years ago. A short, stooped brute shuffles across a cold, windswept landscape, dragging the scavenged leg from a reindeer, and ducks his massive bulk under the rock overhang of a cave consumed in shadow. His scraggly hair covers an apelike face with protruding brow ridges and falls on massive hunched shoulders set on a hairy body with thick short legs. He grunts to the females cowering in one corner of the cave, then turns to face a group of young males, pounding his chest and howling. He has reestablished his dominance, for the short term at least.

The males jump on the deer leg, ripping meat from it with their hands and smashing the bones with crudely developed stone tools. Blood drips from their massive jaws. They grunt, belch, and fart. The females sneak scraps thrown aside in the scuffle. There are more grunts, and fights break out. A few chunks of meat clinging to broken bones are heaved to youngsters hidden in the dark recesses of the cave. They gnaw on the bones, sucking and slurping the marrow.

In the bushes outside the cave lies a body, belly bloated. Flies swarm. Maggots crawl out of the eyeballs and undulate along the severed leg and the bloody guts extruding from the exploded stomach of the dead Neanderthal. No symbolic ritual paraphernalia is apparent. The dead are dumped and left to rot, their stench mingling with the remains of carcasses slaughtered by carnivores and stripped bare by scavenging Neanderthals. Close by lies the bleached white skull of a five-year-old child, picked clean by maggots. There is no life beyond the fight for survival, and the strongest live the longest.

Camped in the valley beyond the Neanderthals' cave is a group of *Homo sapiens.* In contrast to the beastlike Neanderthals, they walk upright, swinging their arms in almost military fashion, and converse in sentences. Their hair has been cut; the men are clean-shaven. They have left their wives and children behind in the adjacent valley. The women and children weep and pray in the men's absence but remain confident of their return.

Discussing the strategy for their upcoming assault on the Neanderthals, the men prepare for battle. Their weapons are beautifully shaped spears, bifacial tools of stone and bone.

They have delayed their attack for one day so they can hold a funeral for a young man who died while hunting a mammoth for the camp. Although the captured mammoth was a youngster, the mother did not give up easily. Red ochre is spread in a grave dug for the occasion. Prayers are whispered into the wind; music is played on specially crafted flutes. A few tears fall to the ground. The men hug one another, keenly aware of the loss of one of their family.

As the next day dawns, they embark on their intended slaughter. There is no contest. The *H. sapiens* outsmart the dim-witted Neanderthals with their superior intellect and weaponry. Neanderthal children lie piled among the stabbed and bloodied dead. There are no Neanderthal survivors; all must die.

The men return home victorious. They now compete against one fewer clan of the diminishing Neanderthals. The climate is rapidly deteriorating and food is becoming scarce – there is not enough for all and only the fittest may survive.

So goes the story of the Neanderthals as told by the victors after millennia had passed. However, an accumulation of new evidence unearthed in Europe suggests this view is flawed. Neanderthals, who first appeared in the fossil record about 200,000 years ago, were definitely not hunched, shuffling brutes; their fossil skeletons are remarkably similar to those of *H. sapiens* and their brain size was larger than that of modern humans.

Archaeologists use different types of stone tool manufacture as markers to identify different *Homo* species (e.g., *H. sapiens* versus *H. neanderthalensis*). They have long considered the beads, bone carvings, and stone tools of the Aurignacian and Châtelperronian from western Europe, dating to the Upper Palaeolithic, to have been made by behaviorally modern humans (*H. sapiens*). Neanderthals are believed to have contributed the older Mousterian stone tool industry of the

Middle Palaeolithic.[1] Some researchers suggest a key explanation for this divergence is that Neanderthals were cognitively inferior; they lacked language and could not have developed the more advanced Châtelperronian tools without input from modern humans.[2] However, the discovery of 36,000-year-old Neanderthal skeletal remains at Saint-Césaire, France alongside Châtelperronian stone tools cast doubt on archaeologists' interpretation that these tools were made solely by modern humans. The Saint-Césaire find suggests they may have been made by Neanderthals – and that Neanderthals may have survived for thousands of years after *H. sapiens* arrived in Europe, co-existing with the newcomer for at least 10,000 years before the last Neanderthal went extinct in southern Spain.[3]

Although archaeologists typically base their understanding of what a species is on its tool kit, it is important to remember that other factors influence tool size and shape. Access to and availability of raw materials, distribution of food resources (which is heavily dependent on changing climate), and group size and composition (larger populations tend to be more culturally innovative than smaller ones) all play a role in tool formation and variation. We know that the climate between 60,000 and 30,000 years ago was highly unstable. Although there was more extensive needle-leaf tree coverage in Europe at that time than there is today, climate simulations indicate the extent and productivity of available human habitats were shrinking.[4] Food resources available to Neanderthals and modern humans would also have been dwindling, and this would have encouraged the competing *Homos* to create new technologies to draw more food from the land.[5]

Wooden javelin-like hunting spears have been uncovered among the remains of butchered horses by a lake in Schöningen, Germany. Each spear was made from the trunk of a thirty-year-old spruce tree. The spear tips were made with wood from the base of the tree, where the wood is the hardest, and the spears' proportions are the same as those of modern javelins, with the center of gravity one-third of the way down the shaft. Their makers, probably *H. heidelbergensis*, created these spears 400,000 years ago. The creation of such sophisticated spears implies a behavioral repertoire that included foresight, planning, and the ability to choose appropriate technology and use systematic hunting skills. The discovery of these spears has caused us to pause and reflect on our previously limited interpretation of early *Homo* and Neanderthal capabilities.

Archaeologists have found few examples of art, symbolism (like burials and fertility symbols), and personal adornment before the appearance of modern humans, which has led them to suggest that such activities are linked to a cognitive requirement. In other words, modern humans were capable of more advanced cultural characteristics, whereas Neanderthals were not.[6] However, Neanderthal burial sites have been found, including a grave for a young male adorned with flowers and red ochre. This suggests the Neanderthals had some capacity for compassion, spirituality, and symbolism. Further, it may mean that this cultural shift in the archaeological record was not related to a specific species, or, at least, not one yet found in the archaeological record.

Although questions remain about whether modern humans and Neanderthals interbred and produced fertile offspring, until recently there has been little in human genetics to suggest a Neanderthal presence. However, in May 2010, new research was published that compared DNA recovered from fossil bones of three Neanderthals with that of five present-day humans from different parts of the world. Researchers found that between 1 percent and 4 percent of the present-day human genome is similar to the Neanderthals. Genetic similarities have been found in genes involved in metabolism and in cognitive and skeletal development. The similarities have been found only in non-African populations, which suggests that potential interbreeding between Neanderthals and *H. sapiens* happened about 45,000 years ago in the Middle East, after modern humans left Africa but before they split into the different populations now found in Europe, China, Papua New Guinea, and elsewhere.[7]

At least one part of the traditional Neanderthal myth is true. Between 50,000 and 30,000 years ago, *H. sapiens* experienced a cultural renaissance in Europe, and Neanderthals, although certainly capable of at least some degree of creative thought, suffered a population decline. The climate was changing, requiring increasingly rapid behavioral adjustments. *H. sapiens* adjusted. Neanderthals disappeared. Some anthropologists believe that competition from the dominant *H. sapiens* combined with a changing climate caused the extinction of the Neanderthals.[8] But did *H. sapiens* actually kill off our Neanderthal cousins?

The human tendency to overkill – intentionally killing more animals or plants than needed for food – may have exacerbated the impacts of catastrophic climate events. Some researchers believe *H. sapiens* overhunted the megafauna (animals weighing more than 45 kilograms) in the Americas, causing or at least contributing to their extinction about 11,000 years ago.[9] The demise of several forest-dwelling marsupials in New Guinea, and the megafauna extinctions in Australia, coincided with, or closely followed, the arrival of over-zealous *H. sapiens* hunters about 46,000 years ago.[10]

To kill for survival is an attribute common to many carnivores, including early *Homo* species, but humans (and wolves, known to slaughter numerous sheep in a pen and then eat a few and leave many of the dead untouched) appear to have developed the capacity to kill for reasons other than survival, as the many racial, religious, and political conflicts in our history demonstrate. Still, it is hard to imagine why humans would choose to decimate the species upon which we rely for survival or, alternatively, a cousin species with which we shared so many characteristics. It may be that humans developed a mean streak that permits us to disassociate from other species and kill them, even if those species are similar to us in all but a few respects. Maybe we lack the ability to see the consequences of our actions or to understand our interdependent relationship with Earth and other creatures, a relationship that has become more difficult to recognize as we move into metropolises far away from the natural sources of our food and material wealth, where our opportunities to connect with nature are limited. A contributing factor is our belief that we are, in fact, the wise man that our Latin name suggests, and that we therefore have the right to dominate all other species. Perhaps this is why we seek control and dominion rather than connection. Yet this dominance, and our fear of being jostled from it, blinds us to our dependence on our environment, other species, and other humans.

Do we justify our actions because we feel we are too intelligent to make mistakes? Has our capacity to dominate and destroy, which frequently exceeds that essential for survival, become a craving dangerous to other species on Earth and to ourselves?

Our enigmatic Neanderthal cousins are now gone. *H. sapiens,* the wise ones, are now common, most likely because we were more able to adjust to a rapidly changing environment. However, with the onset of predicted rapid future climate change and all that it entails, we may experience the same fate as Neanderthals if we continue to resist change and insist on dominating, controlling, and destroying the very environment and species on which we depend for our survival. Perhaps we would be wise to learn a lesson from yet another extinct cousin – *H. floresiensis.* A remarkable recent archaeological discovery tells us more about this previously unknown *Homo* species, affectionately termed "the hobbit," which disappeared around the time humans were spreading across the Americas.

4

The End of *Homo* Diversity

> In the pre-agricultural world, human beings and animals were all hunter-gatherers: they were in a real sense kin. As a child of Bear or Eagle, moreover, a man could hope to reach the parts of his own nature to which he didn't have access: courage, strength, wisdom, freedom, and beauty.
>
> *Carol Lee Flinders,* Rebalancing the World

In 2004, archaeologists made a spectacular discovery in a cave on the small, 354-kilometer-long volcanic island of Flores, east of Java in Indonesia. They found the remains of six tiny people, shorter than human pygmies by 0.5 to 0.6 meters. One female stood only 1 meter tall and weighed just 28.7 kilograms when she died at about the age of 30 years.

These "hobbits," as they were nicknamed, walked on two legs and hunted the dwarf *Stegodon* (a now extinct elephantlike mammal), as well as giant rats and Komodo dragons that today grow up to 3 meters long. The little people lived on their island refuge for at least 78,000 years (from about 95,000 to 17,000 years ago). A cooling climate had lowered sea level to 55 meters below what it is today, but even when sea level fell to its lowest point during the last ice age, about 21,500 years ago, there was still a 24-kilometer sea crossing to the island. Other land mammals were unable to make the crossing.

So how and why did the "hopeful hobbits" go to the island?

A changing climate may have forced the little people to leave their original home when food resources were no longer sufficient. Caught in a storm, or cast adrift on a makeshift raft, they may have drifted for days before landing on the beaches of their isolated new home.

Another possibility is that the hobbits were descendants of *Homo erectus*, the brave and adventurous hominin who left Africa over a million years ago, migrated across Asia, and managed to find a way to cross the open water, arriving on the island of Flores. Early stone tools and fossils dating to this time have been discovered on the island and attributed to either *H. erectus* or *H. habilis*.[1] Some scientists propose that the little people became dwarfs as a result of living on an isolated island. Others suggest that they arrived on the island small-bodied.

New research has determined that the stone tools found on Flores Island are in fact 1 million years old and, further, that *H. floresiensis* most closely resembles *H. habilis*, who lived in Africa 2.4 million years ago.[2] But wherever and whenever they became small, they did not dwarf in the regular sense. Unlike human dwarfs, the hobbits had brains that shrank proportionally with the rest of their bodies, resulting in tiny brains of about 380 cc – the size of an orange. This is smaller than the previously accepted range for *Homo* species and is about the same size as the brain of *Australopithecus*, who lived about 2 million years ago. *H. floresiensis*'s brains were less than one-third the size of our modern brains, which are 1,300 to 1,400 cc.

Since at least 1861, when Paul Broca concluded that, "in general, the brain is larger in mature adults than in the elderly, in men than in women, in eminent men than in men of mediocre talent, in superior races than in inferior races," and "other things equal, there is a remarkable relationship between the development of intelligence and the volume of the brain,"[3] anthropologists and laypeople have argued that there is a strong positive correlation between human brain size and intelligence.

This belief in the connection between brain size and intelligence has persisted despite evidence to the contrary and despite evidence that parts of some early brains were bigger than the corresponding area in the brains of *H. sapiens*. For example, upon finding that fossil skulls of the early European Cro-Magnon possessed a brain size capacity that exceeded that of modern Frenchmen, Broca reasoned that the Cro-Magnon might have larger cranial capacity, but it did not occur in the anterior or frontal parts of the brain, which are responsible for speech, elaborate thought, and voluntary motor activity. Rather, it was in the occipital lobes located at the back of the brain, which are associated with more mundane and primitive activities. Thus, Broca concluded, the Cro-Magnons were incapable of intellectual advancement, let alone developing a civilization. Such an attitude is evidence of our survival of the fittest belief. Even when it appears that an extinct *Homo* may be fitter in at least some capacity than us, we view it as an exception or discount the evidence.

Today, we may think our brains, with larger anterior regions, indicate we are smarter and therefore fitter than Cro-Magnons, Neanderthals, or "hobbits," but consider how successful a modern city dweller would be at avoiding a Komodo dragon by using a stone-tipped throwing spear. The small-brained *H. floresiensis* created stone tools, including a specialized tool kit associated solely with hunting the dwarf elephant *Stegodon*. They retained many of the functional capabilities of *H. erectus* or *H. habilis* and may even have possessed some of the abilities of *H. sapiens*, including the capacity to make stone tools, hunt alone and in groups, make the crossing to an isolated island, and survive in a limited habitat. Further, it was during the time that *H. floresiensis* was living on Flores Island that the so-called revolution of modern behavior occurred in Europe (between about 50,000 and 30,000 years ago) and that aboriginal peoples colonized the New World, including Australia (about 50,000 years ago) and the Americas (around 20,000 years ago).

Yet *H. floresiensis* arrived on Flores before *H. sapiens* reached Australia. They, or their predecessors, were able to cross the sea and colonize Flores Island during the Pleistocene or earlier, at a time when other land mammals were unable to make the crossings, even though sea levels were lower during glacial periods. These small-brained hominins also managed to survive on their little island alongside the Komodo dragon without being annihilated by or exterminating the species on which they depended for survival. And they flourished, even though *H. sapiens* lived nearby for between 18,000 and 38,000 years. And we all know how destructive *H. sapiens* can be – just consider our extinct ancestors the Neanderthals. Some scientists suggest that *H. floresiensis* and *H. sapiens* were kissing cousins – sexual reproduction between the two species was possible.[4]

In the end, the demise of *H. floresiensis* and the *Stegodon* occurred at the same time; a volcanic eruption is thought to have destabilized what must have been a fragile equilibrium, and both vanished.[5] Yet some people believe the hobbits' disappearance was the work of mean-spirited *H. sapiens*, who either destroyed them or the resources on which they depended on the island. Others speculate that the hobbits may yet live as the Indonesian Orang Pendek.[6] However, at this point, tiny bones and skulls buried alongside giant dragon bones and scattered stone tools are all that are currently known to remain of the last of our closest relatives.

H. floresiensis illuminates the importance of our codependence on other species and, yet again, refutes the notion that bigger is better. In our modern, expanding, industrial, and commercialized economy, how many big-brained *H. sapiens* are waking up to the need to adjust our behavior in response to the onset of rapid global climate change, which could eventually cause the extinction of species we rely on for survival? How many are contemplating the importance of environmental and social interaction within and between species and pondering the future of the last remaining *Homo* species - and all other species?

With all our cousin *Homo* species gone, it remained for *H. sapiens* to migrate around and inhabit the world. The following section delves into the incredible geological, climatological, archaeological, and aboriginal histories of the peopling of the world.

Climate and Human Migration

5

Climate and Human Migration

> Man, who even now finds scarce breathing room on this vast globe, cannot retire from the Old World to some yet undiscovered continent, and wait for the slow action of such causes to replace ... the Eden he has wasted.
>
> *George Perkins Marsh,* Man and Nature; or, Physical Geography as Modified by Human Action *(1864)*

Just like *Homo erectus* before them, *Homo sapiens* dispersed and migrated in response to a rapidly changing climate. They did so during what geologists call the last glacial cycle (135,000 to 11,650 years ago), in large measure because of the changes it wrought on their home territories. Geographic barriers like mountains, deserts, and oceans; changing sea level; massive glaciers that soared kilometers in height; sea ice that stretched across the northern and southern oceans; rapidly changing temperatures; rain, snow, sleet, and drought; predators and changes in the availability of plants and animals they relied on for food – all influenced where, how, and when early people lived and migrated. This chapter and the next trace this migration throughout the world as humans searched for new places to live as an onslaught of climate crises struck their traditional territories.

Climate during the last glacial cycle

The last glacial cycle covers a geologic period between 135,000 and 11,650 years ago, when rapid changes in climate had profound impacts on the plants and animals that inhabited Earth. During this cycle, Earth experienced three major cold events, called stadials (these are Marine Isotope stage 6, 4, and 2, abbreviated MIS6, MIS4, and MIS2,

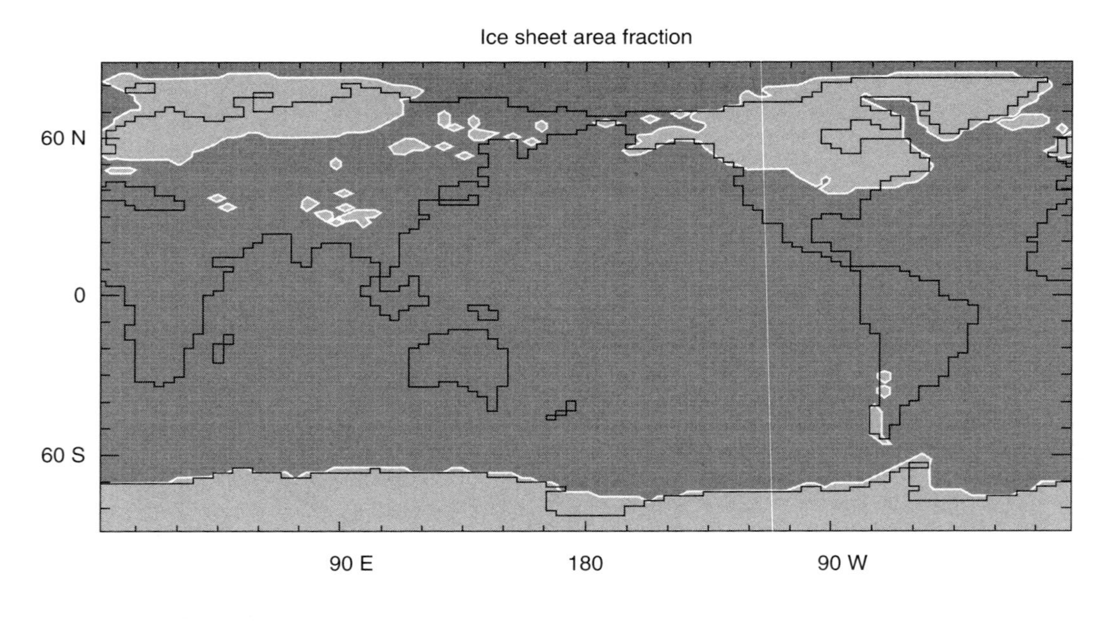

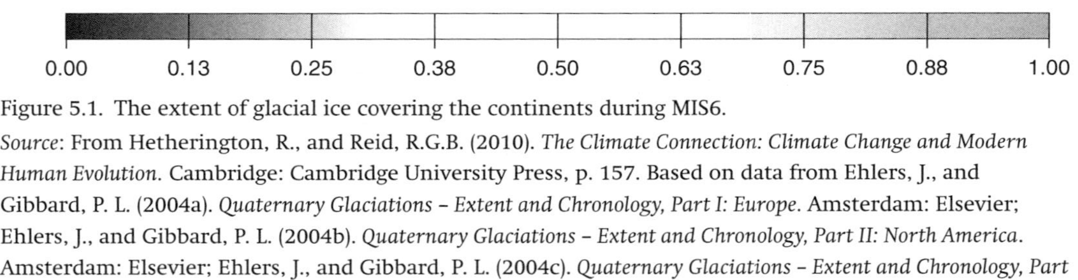

Figure 5.1. The extent of glacial ice covering the continents during MIS6.

Source: From Hetherington, R., and Reid, R.G.B. (2010). *The Climate Connection: Climate Change and Modern Human Evolution*. Cambridge: Cambridge University Press, p. 157. Based on data from Ehlers, J., and Gibbard, P. L. (2004a). *Quaternary Glaciations – Extent and Chronology, Part I: Europe*. Amsterdam: Elsevier; Ehlers, J., and Gibbard, P. L. (2004b). *Quaternary Glaciations – Extent and Chronology, Part II: North America*. Amsterdam: Elsevier; Ehlers, J., and Gibbard, P. L. (2004c). *Quaternary Glaciations – Extent and Chronology, Part III: South America, Asia, Africa, Australia, Antarctica*. Amsterdam: Elsevier.

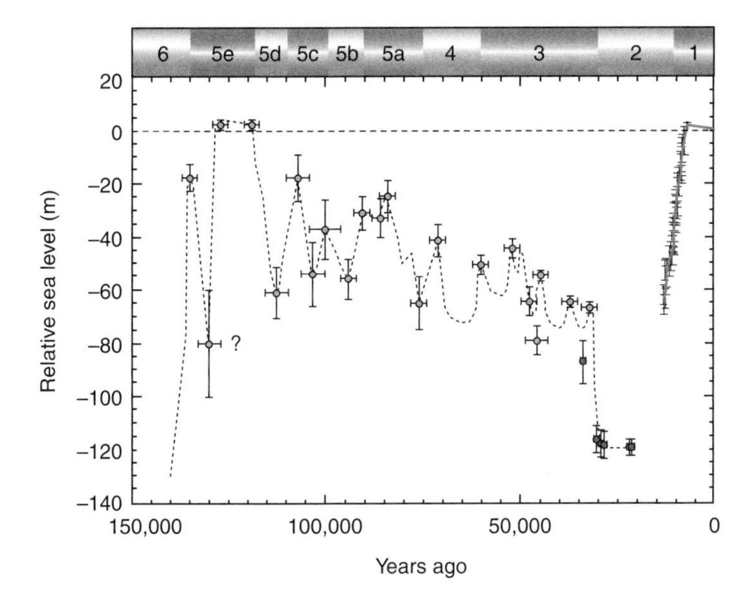

Figure 5.2. The relative sea-level curve for the last glacial cycle. Bars on each sea-level data point indicate the upper and lower limits of error that should be associated with each data point. The main marine isotope stages MIS6 to MIS1, including the substages of MIS5, are shown. Zero indicates present-day sea level.

Source: Lambeck, K., Esat, T. M., and Potter, E.-K. (2002). Links between climate and sea levels for the past three million years. *Nature*, 419, 199–206. Reprinted with permission from Macmillan Publishers Ltd., copyright 2005.

respectively), and two major warm events, called interstadials (MIS5 and MIS3), as well as a number of smaller climate variations. The temperature changes associated with these events, along with the smaller temperature variations, are shown in Figure 1.3.

During the stadials, massive ice sheets extended across the northern hemisphere (see Figure 5.1). Adjacent to some of these, particularly in Eurasia, barren glacial landscapes and polar deserts formed. When glaciers were most extensive, the surface-air temperatures over Antarctica reached as low as 9°C colder than today, global average sea level dropped to 120 meters lower than today (see Figure 5.2), and many of the continental shelves that are now submerged along the edges of world's continents emerged from beneath the sea.

In contrast, during interstadials, many of the world's glaciers disappeared, and warm, wetter conditions encouraged extensive forests to develop; plants grew in areas of Africa now covered with desert. Surface-air temperatures over Antarctica reached as much as 2°C warmer than today, and global average sea level rose to as high as 6 meters above today's levels, drowning many of the coastal regions where many people currently live and submerging the continental shelves once again.

How did a changing climate affect the home of early *H. sapiens*?

Throughout the last glacial cycle, plant and animal populations and the environments they lived in expanded, contracted, and altered as climate changed. Early humans shifted their territories, sometimes following the migration of large mammals that traveled in search of more abundant food and water. Although drops in temperatures during the cold intervals were not as significant in the tropical regions as they were in the higher latitudes like Europe and northern North America, the climate of Africa – the birthplace of *H. sapiens* – was still affected. For example, during the height of the last ice age, less rain fell in some regions of interior Africa; as a result, deserts expanded.[1] The monsoon in south and east Asia weakened, causing a desert barrier to form between East Africa and south Asia that is still evident today.[2] Yet on the newly exposed continental shelves, freshwater springs developed, providing productive oases with prolific vegetation and plenty of nutritional shellfish.[3] These exposed continental shelves provided a haven from desiccated interior regions.[4]

H. sapiens living in areas where rainfall was dropping and deserts were expanding would have felt the need to move to "greener pastures." Other phenomena may also have forced them to move, including population increases, diseases transmitted by plants or other animals, or deadly microbes unwittingly carried by other people.[5] Although these challenges were difficult, they also helped our human ancestors improve their survival skills and develop the ability to live in a variety of different environments. Using their newly acquired skills, *H. sapiens* embarked on a series of migrations around the world.

When *H. sapiens* first emerged, were they just like us?

The archaeological sites where the oldest *H. sapiens* fossil bones have been unearthed are in Africa. They are 195,000 years old in Omo, Ethiopia, and between 130,000 and 110,000 years old along the eastern edge of what is now Lake Victoria in East Africa and at Mumba, Tanzania. It is no coincidence that these regions spawned the morphological, physiological, and behavioral characteristic changes that resulted in the emergence of *H. sapiens*; they experienced some of the most extreme climate volatility in Africa. This volatility stimulated a series of events and crises that resulted in the emergence of modern humans.

Early *H. sapiens* were physically like us, but they did not possess the behavioral qualities that we now think of as "modern." There is no evidence to suggest that the human brain has changed in its anatomical form since the first *H. sapiens* originated 195,000 years ago. At that time, the brain had the same potential to develop thoughts, ideas, communication, spirituality, artistry, and technical complexity as our modern brains do today; however, it took a combination of both favorable and stressful environmental conditions, combined with a complex social environment, to bring out the innovative potential that we associate with modern *H. sapiens.* Together with Robert Reid (2010) I suggest that what allowed us to realize the potential to become behaviorally modern was *crisis* (especially climatic), *communication* between previously isolated groups, and the *collaboration* of these groups in solving problems. When these circumstances were in place, revolutionary change was stimulated. Migration was an important part of this process, as well as being the precursor to what became the "modern" human condition, first seen in Africa about 70,000 years ago and in Europe between about 50,000 and 40,000 years ago.[6]

When and why did *H. sapiens* migrate out of Africa?

When deserts began to expand in interior Africa, *H. sapiens* moved to the exposed continental shelves. Around 164,000 years ago, when sea levels were 130 meters lower than they are today, early humans took advantage of what apparently were plentiful marine resources at Pinnacle Point on the south coast of Africa. Archaeologists have also found artifacts dating to 125,000 years ago buried along the Red Sea coast of Eritrea, an enclosed basin between Africa and Asia. It appears that *H. sapiens* survived along this coast by eating marine food just prior to a rapid sea-level rise 130,000 years ago. *H. sapiens* also migrated to the southern tip of Africa at Klasies River Mouth, South Africa, 118,000 years ago and again between 105,000 and 94,000 years ago.

By 110,000 years ago, conditions in Africa began to seriously deteriorate, and the habitable regions where *H. sapiens* were living were shrinking in size. Many of our early relatives were living near Lake Victoria, East Africa, and the volatile climate in this region, combined with the concentration of human populations, is thought to have stimulated a rapid evolution in *H. sapiens.* Communication between different individuals and groups of individuals increased, as did the exchange of novel ideas focusing on how to deal with the change they were experiencing. As a result, *H. sapiens* came up with new innovations and technologies that helped them survive. Some archaeologists believe that the stressful conditions were enough to inspire the transition from the African Middle to Late Stone Age. These new developments are also thought to have been the key innovations that allowed *H. sapiens* to migrate out of Africa.

By 100,000 years ago, a lessening of rainfall brought prolonged periods of drought to central Africa; however, the eastern coastal area of central Africa, North Africa, and the Middle East were receiving more rainfall, which made them more attractive to *H. sapiens.* The desert, which had previously acted as a barrier to human movement into North Africa, retreated as grasslands expanded. *H. sapiens* migrated to the Levant, into what is now Israel (see Figure 5.3), where deserts were retreating and plants and animals that were not significantly different than those in North Africa had also migrated. The first *H. sapiens* fossils discovered outside Africa were uncovered here and dated to between 119,000 and 85,000 years ago.

Why did humans beat a hasty retreat back to Africa?

Between 90,000 and 75,000 years ago (MIS5a), global average surface-air temperatures were about 2°C colder than they are today, and global sea level ranged from between 65 meters and 25 meters lower than today. The polar desert south of the northern European ice sheets grew to extend across much of central Eurasia and likely blocked the migration of *H. sapiens* northward out of the Middle East. At the same time, the Levant desert again began to expand, and *H. sapiens*, facing a shrinking habitat, beat a retreat back to Africa. Archaeologists have found that the number of *H. sapiens* sites in Africa that date to between 90,000 and 60,000 years ago increased, suggesting a growing *H. sapiens* population in Africa. Yet the majority of *H. sapiens* archaeological sites have not been found in the interior of Africa, probably because deserts there were expanding and rainforests were shrinking, making the areas where early humans traditionally lived less habitable.

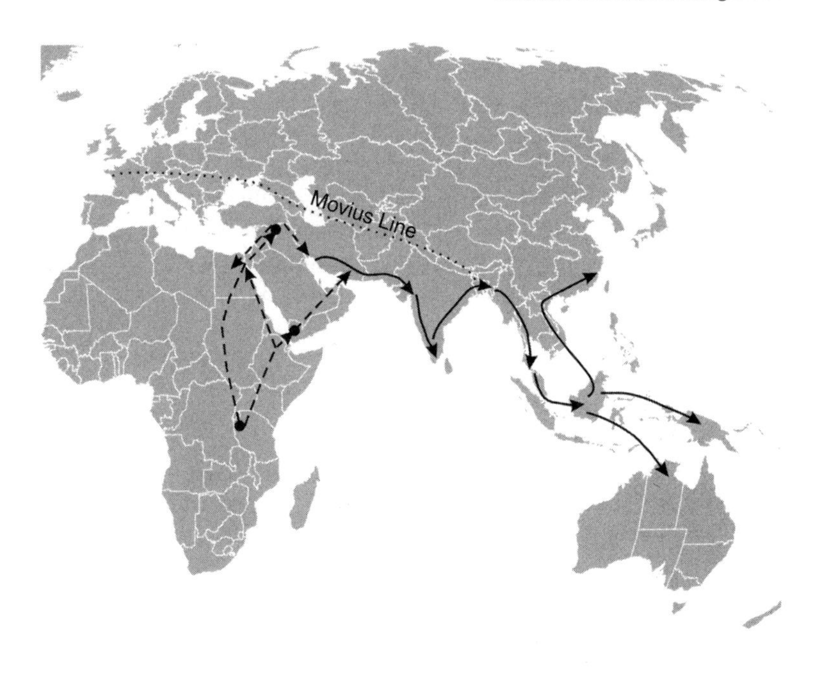

Figure 5.3. Hypothetical routes out of Africa, highlighting a southern coastal route along the Indian Ocean coastline that may have been taken by humans migrating out of Africa. Also shown is the theoretical "Movius line" first proposed by archaeologist H. L. Movius, east of which has yet to be found the teardrop-shaped hand axe – the signature tool of the Middle Palaeolithic Acheulean stone tool kit.

Source: From Hetherington, R., and Reid, R. G. B. (2010). *The Climate Connection: Climate Change and Modern Human Evolution.* Cambridge: Cambridge University Press, p. 90. Based largely on Forster, P. (2004). Ice ages and the mitochondrial DNA chronology of human dispersals: A review. *Philosophical Transactions of the Royal Society of London, Series B, Biological Sciences*, 359, 255–64; Forster, P., and Matsumura, S. (2005). Did early humans go north or south? *Science*, 308, 965–6; and Oppenheimer, S. (2003). *Journey of mankind.* Interactive trail adapted from Out of Eden/ The Real Eve, www.bradshawfoundation.com/journey/

Instead, two-thirds of the sites have been found in coastal regions, where sea level ranged between 25 meters and 65 meters lower than today.[7] The shrinking human habitat inside Africa encouraged those living in interior regions to migrate to the coast, where the climate was more equable.

A glacial climate and a volcanic eruption play havoc with *Homo* species

By 75,000 years ago, the world had fallen into a cold, full glacial interval that lasted until about 60,000 years ago. Temperatures in places reached as much as 9°C colder than today over Vostok, Antarctica.[8] Sea levels fell, exposing the continental shelf. *H. sapiens*, who had moved to this newly exposed and relatively productive African shelf between 90,000 and 60,000 years ago, migrated east along the coast toward Southeast Asia.

At the same time, to the north in central Eurasia, advancing ice sheets and an expanding polar desert forced Neanderthals to seek refuge in the southwesternmost regions of Europe where the environment was less hostile.

Meanwhile, 71,000 years ago in Indonesia, where *H. floresiensis* had lived for 24,000 years, a huge volcano exploded. The eruption of Mount Toba, in northern Sumatra, probably caused the already glacial global climate to cool even more. Plants and animals were affected, their growth and productivity impaired.

These conditions had a severe impact on the size of the global *H. sapiens* population, which remained at about 600,000 individuals until shortly after 40,000 years ago. Those that survived had to rapidly change their behavior to cope with what some say was a volcanic winter caused by Mount Toba's eruption.[9] Incredibly, *H. floresiensis* managed to survive this catastrophe, probably by eating giant rats, a now extinct elephant-like mammal (the dwarf *Stegodon*), and possibly Komodo dragons. *H. floresiensis* remained on Flores Island until about 17,000 years ago, when another volcanic eruption is believed to have caused both it and the *Stegodon* to vanish. Modern humans did not arrive on the island until about 11,000 years ago.[10]

Another move out of Africa – where and why?

Scientists studying the mitochondrial DNA of early humans have found new glimpses of where modern humans came from and how they migrated around the world. Mitochondrial DNA is the DNA of the energy-making mitochondria found within human cells. It contains a small fraction of our total DNA and is passed from a mother to her offspring. By researching mitochondrial DNA, scientists have come to believe that *H. sapiens* dispersed out of Africa between 80,000 and 60,000 years ago and then rapidly expanded into Asia and, in a much more restricted way, into colder northern environments.[11] According to this research, behaviorally modern humans left Africa around 65,000 years ago, following a route along the coast of the Indian Ocean and into Southeast Asia before reaching Australia around 50,000 years ago (see Figure 5.3).[12]

Other scientists studying simulations of the world's climate and vegetation during the last glacial cycle have observed a huge drop in the productivity of global vegetation 60,000 years ago.[13] This would have meant a large decline in available food for early humans and may explain why the number of *H. sapiens* archaeological sites in Africa fell drastically during this time.

A changing climate 60,000 years ago likely forced early humans into a restricted number of habitable regions as food resources dwindled. Many of these regions, where food resources remained relatively plentiful, were located near water. Living so close to one another would have meant that individuals with different backgrounds and experiences interacted more frequently with one another. Evidence of a 60,000-year-old Neanderthal skeleton at Kebara 2, Israel, has led some archaeologists to suggest that even Neanderthals coexisted with modern humans in this region.

There is no doubt that during this cold, dry, glacial interval it became more difficult to find food. This, combined with the concentration of people in a limited number of territories, meant humans needed to develop new technologies if they were going to survive. And these resourceful humans did indeed create new tools that improved their ability to find food, which in turn allowed them to move into and survive in new environments. The new Late Stone Age technology was developed in Africa, and an abrupt transition from Middle to Upper Palaeolithic stone tools occurred in Europe. Some researchers consider this latter development to reflect a "revolution in modern human behavior" (see Figure 2.2). What also followed was the growth and expansion of the human population; by 35,000 years ago, the global human population had reached 4 million.

H. sapiens reach Australia and megafauna go extinct

A conservative estimate puts the first *H. sapiens* in Australia between about 50,000 and 40,000 years ago.[14] At the same time, or shortly thereafter, a large megafauna extinction occurred. All 19 Australian marsupials exceeding 100 kilograms, including a marsupial the size of a hippopotamus, a 200-kilogram kangaroo that grew up to 3 meters tall, and a 130-kilogram marsupial lion, went extinct. In addition, 22 of the 38 marsupial species between 10 kilograms and 100 kilograms, three large reptiles, and a giant flightless bird all suddenly disappeared. Many people believe this was caused by humans; however, the 1.8-meter-tall flightless bird *Genyornis newtoni*, which fed primarily or exclusively on a specific type of grasses (C3 grasses), also went extinct at a time when those grasses were dying off and being replaced by a different type of grasses (C4 grasses) that it could not eat. At the same time, the flightless emu, *Dromaius novaehollandiae*, which had a broader diet that included C4 grasses, survived. It was able to alter its diet as the plants that grew in its home territory changed.[15] As a result of these findings, scientists now think that climate, along with humans, likely played a role in the extinction of the flightless bird *Genyornis newtoni* and perhaps other megafauna species as well.

Why did it take so long for *H. sapiens* to reach Europe?

One of the great puzzles in the migration history of *H. sapiens* has been how humans managed to get all the way to Southeast Asia and Australia before arriving in Europe. Why didn't our modern human ancestors just head north out of Africa and populate Europe? After all, Europe is a lot closer.

Europe wasn't always barren of *Homo* species. In fact, 500,000 years ago, an ancient relative, which some archaeologists call *H. heidelbergensis* and others have named "archaic *H. sapiens*" (meaning a species intermediate between *H. erectus* and behaviorally modern *H. sapiens*), found its way to Boxgrove in England and Mauer, Germany. *H. heidelbergensis* likely traveled to England when sea levels much lower than today's exposed the shelf that links England with northern Europe. Yet even though *H. heidelbergensis* managed to migrate to England half a million years ago, it took modern humans another 450,000 years or more before they reached Europe. Why?

The reason may be that during much of the last glacial cycle, after *H. sapiens* first appeared, huge ice sheets extended across much of northern Europe and Asia, and these, combined with an expanding polar desert, blocked the route that early humans would have used to migrate from Africa to Europe. These prohibitive conditions probably also forced Neanderthals, who were already living in Europe, to find refuge in a relatively small habitable region in the southwesternmost region of Europe. Another desert barrier in the Middle East, which for thousands of years has acted as the crossroads between Africa, Asia, and Europe, may also have blocked *H. sapiens* from moving north out of Africa and into Europe.[16]

It wasn't until 41,500 years ago that modern humans first arrived in Europe; the oldest *H. sapiens* fossils in Europe have been uncovered in Germany and Romania. Archaeologists have also found European evidence, from the same time, of the development of new stone tools, called the Upper Palaeolithic stone tool technology. These bone, antler, and ivory tools, some containing notations, are believed to have been made by *H. sapiens*. The toolmakers made, and adorned themselves with, pendants and perforated marine shells, and created ivory and stone beads along with abstract and naturalistic art. They also developed elaborate distribution systems. Some believe this signals a revolution in modern human behavior, one that helped *H. sapiens* survive what was fast becoming a worsening glacial environment. *H. sapiens* were also able to successfully compete with the neighboring Neanderthals for a limited number of resources.

Interestingly, many of the same characteristics that are typical of the Upper Palaeolithic stone tools in Europe individually appear earlier and much more gradually in the South African archaeological record during the transition from the Middle to Late Stone Age, beginning as early as 250,000 to 300,000 years ago, but most strikingly between 70,000 and 60,000 years ago.[17] Yet it is clear that, although individual new discoveries occurred over a long time and were each revolutionary in themselves, the coming together of all these new behaviors simultaneously in Europe 40,000 years ago produced nothing less than a revolution. *H. sapiens* stopped behaving archaically and developed the spark of creativity that resulted in what we now identify as modern. These modern advancements set the stage for humans' migration into the New World.

6

Braving the New World

...some started over the glacier.
These are the ones who came down the Chilkat, the relatives of my fathers,
The Dakl' aweidí.

Excerpt from the Tlingit story Kák'w Shaadaax' x'éidáx sh kalneek

Perhaps the greatest mystery remaining in the migration history of modern humans is how and when early people appeared in the Americas. Archaeologists believe they arrived some time during the last ice age (between 30,000 and 11,650 years ago), when massive ice sheets covered much of the northern hemisphere. Many think they arrived around or soon after 21,500 years ago, at the time when glaciers reached their greatest extent.

What was the environment of the Americas like during the last ice age?

The environment at that time was extreme! Not only did huge ice sheets, some 2.5 kilometers thick, cover the northern half of North America, but coastal areas next to the ice sheets were drowned, whereas offshore, in areas where ice was thin or absent, the continental shelf bulged upward, exposing large coastal plains.[1]

Although Alaska, in the far northwest corner of North America, is now seen as a cold, snowy place, it was mostly free of ice during the last ice age. Only southern Alaska was ice-covered, and little snow fell elsewhere. As a result, grizzly bears and other species of mammals survived there during the last ice age. Conditions were similar in much of the Arctic, where very little ice covered dry, sage grasslands and shrub tundra.

Beringia – the huge continental shelf that extends from northeastern Siberia to the Mackenzie River in the Yukon of Canada (Figure 6.1) – was exposed during the last ice age when sea level dropped to as low as 120 meters below today's levels. Massive ice-age mammoths ranged across the productive herb-tundra of Beringia's immense coastal plain.[2] To the west, in Asia, a huge polar desert, barren of vegetation, formed where little to no ice existed. Nearby on Wrangel Island, off the northeastern coast of Siberia, dwarf mammoths lived until long after their huge cousins went extinct.

Along the edges of the colossal North American ice sheets, glacial lakes formed beside cold grasslands. Two large lakes, Bonneville and Lahontan, extended across the American southwest, and the enormous glacial Lake Agassiz stretched over a million square kilometers at the head of the Laurentide ice sheet in eastern Canada.

On Canada's west coast, ice sheets that were kilometers thick depressed the mainland, causing the sea to inundate land that today is 120 meters above sea level and covered in rainforests. At the same time, just 50 kilometers to the west, the ocean floor that is now drowned under 120 meters of seawater became dry land. Little to no ice covered this new land, which rose upward much as the edge of a waterbed does when your body pushes down its centre. New coastlines formed with bays that hosted abundant shellfish and marine mammals. This is where some archaeologists believe early migrants to North America thrived.

When a warming climate finally caused the ice sheets to melt, the land on the continent, which had been pushed down by heavy loads of ice, sprang back up. At the same time, global sea levels rose as water from the melting ice sheets rushed back to the oceans. The shelf and islands just off Canada's west coast, which had bulged upward as the weight of the ice sheets pushed the mainland down, collapsed when the load of ice from the mainland disappeared. The sea raced inland. Old coastlines were drowned and new ones appeared, having shifted more than 100 kilometers within the span of a few human lifetimes.

These enormous changes had a tremendous impact on the people living and migrating in this area. Oral histories tell of the challenges they faced. These stories have been passed down through generations. Here is part of a Tlingit oral history from Canada's Pacific Northwest coast.

> One's name was Awastí and the other Koowasíkx,
> these elderly women.

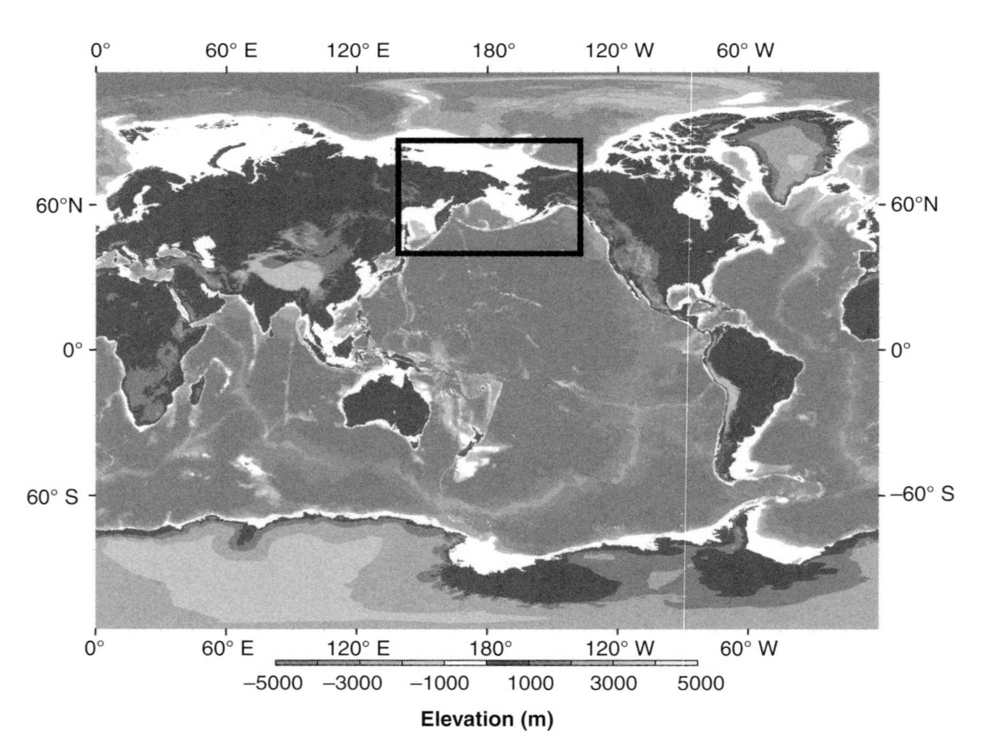

Figure 6.1. Land elevation and ocean depths showing continental shelves of the world and indicating the location of the Bering Strait continental shelf (Beringia).

Source: From Hetherington, R., and Reid, R. G. B. (2010). *The Climate Connection: Climate Change and Modern Human Evolution*. Cambridge: Cambridge University Press, p. 147.

They are the first ones who were pushed under the glacier.
Having drifted under it and through to the other side,
they started singing.
Floating under the glacier
gave them their song.
Based on this
a raft was made.
Some went on it.
Under it, under the glacier, they floated,
down the river.
But many of them
were afraid
to float under the glacier.
This is why they started over it,
some started over the glacier.
These are the ones who came down the Chilkat, the relatives of my fathers,
The Da<u>k</u>l' aweidí.
They became the Chilkats.[3]

How and when did people first come to the Americas?

The Clovis first hypothesis

During the last ice age, mammoth, mastodons, horses, the short-faced bear, the dire wolf, and the American lion ranged the Americas. It is into this new world of exotic animals and ice-age landscapes that the Clovis people migrated. Archaeologists believe the Clovis people came across the subcontinent of Beringia during or near the end of the last ice age. The Clovis people brought with them their beautifully fashioned bifacial spear points, which they used to hunt the ice-age megafauna.

Clovis points were first found near Clovis and Folsom, New Mexico (ergo, the name given to the Clovis points), in the 1920s and 1930s and were later dated to as much as 13,400 years old. Some of the points have been found embedded in buried skeletons of extinct bison. These discoveries led archaeologists to believe that the Clovis people were big-game hunters who followed the massive ice-age mammals across the Beringian landbridge and were the first humans to inhabit the Americas. According to the "Clovis first" hypothesis, the Clovis people migrated southward between the giant ice sheets that stretched across the North American continent, taking advantage of an "ice-free corridor" east of the Canadian Rocky Mountains.

Yet the Clovis culture appears to have been short-lived. Clovis stone tools disappeared about 12,800 years ago in western and central North America and about 12,600 years ago in eastern North America. Archaeologists kept digging for new evidence, but as time passed, flaws began to appear in the Clovis first hypothesis.

A new hypothesis – three migrations, not one?

In the mid-1980s, a new hypothesis for the peopling of the Americas developed. According to this one, there were three migrations of early people who came from Siberia. Based on studies of fossilized human teeth, anthropologists suggested that the oldest people in the Americas were the Paleoindians, a steppe-adapted culture, who brought with them their Clovis tools and spread through North, South, and Central America. The second were the Beringian Paleo-Arctic people, a forest-adapted people from Siberia, who were the ancestors of the Na-Dené speakers of North America. The third came from the Siberian coast and became the Eskimo-Aleut speakers of the extreme north.

As the evidence grew, both of these theories became increasingly questioned. In 1996, two young men discovered a human skull buried along the Columbia River at Kennewick, Washington. Later, more bones were found; a virtually complete skeleton was unearthed from the mud. It was the skeleton of a middle-aged man, nicknamed "Kennewick Man," who archaeologists believed had been hunting or fighting some time during his life, because they found a leaf-shaped stone point buried in his hip bone. He had lived about 9,500 years ago. Interestingly, the shape of his skull, cheekbones, nose, and chin were not typical of those found in Native Americans today, but were more similar to those of Caucasian people. His teeth resembled those from people in Polynesia, Southeast Asia, and Micronesia, or of the Jomon-Ainu people of Japan. These features may mean that he had a south Asian origin or, alternatively, that early Americans were more diverse than originally believed.

Other new evidence has been unearthed that has caused many archaeologists to reconsider the Clovis first and the three-migrations hypotheses. Along the ancient Kamchatka River in eastern Siberia, Nikolai Dikov unearthed the house floors for more than twenty roughly circular wood-framed skin homes in 1961. The stone-encircled hearths found in the center of the dwellings at what became known as the Ushki Lake archaeological site contained charcoal and fish bones. In the deepest and oldest layer of this site, more dwellings were found, the largest of which was 100 square meters. Also found were stone beads, stone arrow points, and burial pits; this oldest layer was originally dated to 16,750 years or older. The Ushki Lake site became critical because it was believed to be antecedent to Clovis; this suggested that the Clovis people had originated in western Beringia. However, more recently the site has been redated and is now believed to be 12,900 years old, centuries younger than the oldest Clovis points found in the Americas. The mystery of the origin of the first peoples in the Americas continues.

New sites have been discovered in the Americas that do not contain any Clovis stone tools. Sites like Meadowcroft in Pennsylvania, Topper in South Carolina, Taima-Taima in Venezuela, Alero Tres Arroyos in Tierra del Fuego, and Toca do Boqueiráo da Pedra Furada in northeastern Brazil all appear to be as old as, or older than, Clovis. One in particular, Monte Verde in Chile, has received a great deal of interest and acceptance. A child's footprints were uncovered here, along with animal-skin dwellings and non-Clovis stone tools. Seaweed samples at the site date to 14,200 years ago, about 800 years older than the oldest Clovis sites yet found. Although we may ultimately find that the dating of some of these sites is wrong, other sites found south of glaciated areas that are older than Clovis, including the Monte Verde site, are difficult to contest.

Does the environment play a role in influencing migration routes?

It is now clear to researchers that any theory about the peopling of the Americas should take into account what influence, whether beneficial or limiting, the environment had on early people. Geologists examining the history of the North American ice sheets are questioning the idea that the Clovis people migrated through an ice-free corridor. This is because the corridor was not open when the Clovis people were supposed to have migrated through it, more than 13,400 years ago. In fact, between about 24,000 and 13,500 years ago, the Laurentide ice sheet that was expanding westward across Canada met and coalesced with the western Cordilleran ice sheet just east of the Canadian Rocky Mountains. These two ice sheets formed a massive ice wall that did not part until 13,500 years ago, which means that early people who lived in sites south of the ice sheets more than 13,500 years ago could not have migrated down to southern North America and South America through an ice-free corridor. Monte Verde, for example, is 700 years older than the time when the corridor first opened; further, it is located at the southern tip of South America, a continent and a half away from the southern edge of the ice sheets. This means that people either came via a different route or arrived in the Americas before 24,000 years ago (i.e., before the two ice sheets converged to form an ice wall).

There were other environmental barriers that influenced how, when, and where early people could migrate into and around the Americas. The melting of the Laurentide ice sheet in eastern Canada generated glacial Lake Agassiz, which extended over about a million square kilometers. It flooded much of what is now Ontario, Manitoba, and parts of Saskatchewan and the United States, and made colonization of these areas virtually impossible. Glacial Lake McConnell, which formed east of the Mackenzie River, would have blocked the movement of people and animals through this region between about 13,400 and 10,200 years ago. Long glacial lakes formed in the valleys of what is now British Columbia. Glacial Lake Missoula captured glacial meltwater in a region now covering parts of Montana, Idaho, and Washington. It formed huge channels in the land every time it burst – which occurred an estimated thirty times! Geological evidence from the last ice age is increasingly being considered by archaeologists in their quest to discover the history of the first peoples of the Americas.

Other theories for the peopling of the Americas

Intriguing and pertinent archaeological evidence found in Bluefish Caves, northern Yukon, further bolsters the idea that people either came across Beringia before the ice sheets blocked the way south, some 24,000 years ago, or came to the Americas via a different route. According to archaeologist Jacques Cinq-Mars and others, it is at the Bluefish Caves that humans with their stone and bone tools survived by eking out a living north of the ice sheets among ice-age horses, mammoths, caribou, birds, and fish. Evidence in the caves is as old as 16,100 years ago. If the dates are correct, these people must have migrated across Beringia earlier than the Clovis first model would suggest, or came via another route.

One such alternate route may have begun along the coast of northeast Asia, crossing or migrating along the coast of Beringia before moving from the north to the south along the Pacific coast of the Americas. Another proposed alternative puts early people in boats voyaging directly to the Americas.

C. J. Heusser, A. D. Krieger, as well as K. Macgowan and J. A. Hester Jr. first proposed the coastal migration route in the early 1960s.[4] In 1979, Knut Fladmark published an extensive analysis of this route.[5] Fladmark was able to ascertain that the reason no early coastal settlements had been found along Canada's west coast was because sea level had risen since the last ice age, drowning those early sites.

Since then, others have sought to find evidence of a coastal migration route. In 1985, after researching the diversity of aboriginal languages in the Americas, R. A. Rogers published evidence that showed much greater aboriginal language diversity in areas south of the ice sheets, with the greatest diversity found along the Pacific Northwest coast.[6] Not long after that, anthropologist Ruth Gruhn discovered 64 aboriginal languages in California, 200 to 350 in Middle America, 1,500 in South America, and 12 along the coast of the Gulf of Mexico.[7] Looking into eastern and central North America, she found only one. The number of aboriginal languages in a given area is believed to directly correlate with the length of time people have lived in the area. The more languages that exist, the longer people have lived there. Ruth Gruhn calculates that humans occupied the Americas for at least 35,000 years and that they originally lived along the coast before moving inland.

What, if anything, do ancient bones and shells tell us about a coastal migration route?

Some of the oldest reliably dated human bones in the Americas have been found along the Pacific coast of North America. In a cave on Prince of Wales Island in Alaska, E. James Dixon unearthed an 11,200-year-old human jaw bone. Also discovered on the island were numerous ancient mammals, including a 41,000-year-old black bear.[8] Just to the south, where the islands of Haida Gwaii were virtually ice-free during the last ice age, the adjacent continental shelf to the east bulged upward, creating a landbridge that connected the islands to the mainland. Seafood could be found along the beaches of the landbridge at least 15,700 years ago.[9] Farther south, on Santa Rosa Island off the southern coast of California, human bones have been dated to 13,400 years ago.[10] Unlike Haida Gwaii, the island was not connected to the mainland during the last ice age, so the human bones imply the earliest use of watercraft in North America. Much farther south, at the coastal site Quebrada Tacahuay in Peru, early people processed marine animals, including birds and fish, 12,500 years ago.[11] These exciting finds support the potential of a coastal migration route, yet they still do not prove it existed.

Were there other possible routes?

As for any good mystery, there are many possibilities and few clues. Other possible routes into the Americas have been suggested. Archaeologists Dennis Stanford and Bruce Bradley think that prehistoric European maritime hunters and fishers migrated by boat, either south of the polar ice that covered Iceland, Greenland, and all but the most southerly parts of Ireland and England, or directly across the Atlantic Ocean from the Iberian Peninsula in Europe. They base this theory, in part, on the similarity between the Solutrean stone tools that appeared in Europe between about 26,500 and 19,500 years ago and the Clovis artifacts found in North America. However, archaeologist Lawrence Guy Straus disagrees.[12] He says the 6,000-year gap between the end of the Solutrean stone tool technology in Europe and the beginning of Clovis in the Americas means these two stone tool technologies were not connected. Further, he suggests the tools are not that similar; many of the Solutrean stone and bone tools found in Europe have not been found in the Americas. Yet the similarity between human skeletons of modern Europeans and that of the 9,500-year-old Kennewick Man, found on the Columbia River in Washington, have increased the profile of this route. A recent study simulating prehistoric transoceanic crossings has shed more light on the possibility of a European source for early Americans. This study found only a 1 percent probability of success when paddling across the ocean from central Europe to the Americas in under 180 days; however, crossings from Europe via Iceland and Greenland had higher probabilities of success (2 to 3 percent) and were much faster (72 days).[13]

Physical anthropologists, who study the similarities between the shapes of skulls among different human groups, believe there were two migrations into the Americas. The first was by Southeast Asians before the last ice age; the second by Mongoloids from northeast Asia. Evidence from human bones found in Lagoa Santa, Brazil, suggests that late and modern Native Americans resemble northeastern Asians, whereas earlier Paleoindian skulls resemble Australians, Melanesians, and Africans.[14] For example, Walter Neves and his colleagues say the skull of a 25- to 30-year-old male, part of a 10,000-year-old skeleton found in a shell midden site in Capelinha, Brazil, along with others from South America, are similar to Paleoindians and to Australians,

Melanesians, and Africans. New discoveries continue to be made across the Americas, from Patagonia and the Colombian Highlands to Mexico and Florida, and the debate is intensifying.

Some suggest a connection with Africa. The earliest known female skeleton from South America, a 12,000-year-old adult female from Lapa Vermelha IV, is strikingly similar to the Dogo, Yeit, and Zulu in Africa. Other archaeologists disagree.[15]

Genetic evidence contributes more intrigue

Geneticists studying mitochondrial DNA and Y-chromosomes suggest people first came to the Americas between 20,000 and 15,000 years ago, during or just after the height of the last ice age.

Researchers looking at mitochondrial DNA, which is passed down from the mother, discovered that Amerindians possess a different type of DNA than do Na-Dené and Eskimo-Aleut. This suggests there were probably at least two migrations. Also fascinating is the discovery that the various groups of human DNA cluster in different geographic areas of the Americas. Group A is prevalent in the north and decreases southward; group B is virtually absent in northern North America and the Andean Indians; groups C and D are much more prevalent in the south and decrease in frequency northward; group X is almost exclusively found in North America among North American Amerindian populations; and Central American populations possess virtually only two groups (A and B).

When looking at Y-chromosomes carried by early fathers to the Americas, scientists found more astonishing insights. One group (haplotype Q-M3) increased in frequency from north to south; another (P-M45) is more predominant in the north. Y-chromosome research also indicates that the original father lived 22,500 years ago in the Yenissey River Basin and Altai Mountains of Siberia. Later, a rival father migrated to the Americas. His Y-chromosomes are found more commonly in the north.[16] The DNA research relating to both early mothers and early fathers suggests more than one migration to the Americas. They also suggest the potential for movement from the south to the north.

When scientists looked at the DNA of the human jaw found on Prince of Wales Island, they found that the mitochondrial DNA was from group D and the Y-chromosomes were from group Q-M3. Both of these groups have a higher frequency in the south and decrease in frequency northward. The lineages are linked with the Chumash tribe in California and the Cayapa coastal tribe in Ecuador. This evidence implies a coastal migration route - and potentially one from the south to the north.[17] The Chumash people made the *tomol* or sewn-plank canoe, found elsewhere only on the coast of Chile and among the Pacific Islanders. These large canoes may have allowed the Chumash people to trade and develop social, political, and economic relations with the Nuu-chah-nulth people of coastal Canada. No one knows whether the *tomol* was invented by the Chumash people or originated in Polynesia. Either way, it suggests a Polynesian-Californian connection.

Could early people have migrated in boats?

New genetic evidence certainly intensifies the mystery and makes us stop and consider migration scenarios that would put early people in southern areas before northern areas. This would necessarily mean that people migrated across the water, a feat that we have typically felt beyond the scope of early people. But if we consider the possibility that early people had boating technology of some type, then how likely is it that they would migrate to the Americas, and what route would they take?

One scenario has them migrating rapidly along the coast - from northeast Asia across Beringia, for instance, and down the Pacific coast of North and South America during the last ice age - then subsequently migrating back northward. Alternatively, they could have migrated across the ocean from other islands or continents at any time during the last ice age. Research into the feasibility of prehistoric transoceanic migrations, done by Montenegro and colleagues, showed that early people in boats could potentially have crossed from Japan to North America in eighty-three days, with up to 8 percent success rate.[18] They could also have crossed from northern Africa to South America in ninety-one days, with more than 13 percent success rate, and from either Scandinavia to Iceland and Greenland or from the Kamchatka Peninsula to North America and Alaska.

Yet these waterborne routes are only feasible if the migrants had the skills and materials to build watercraft. We are aware that some early people used boating technology, but we are far from understanding its origin, diffusion, and the ability of early people to use boats to migrate to the Americas. The mystery persists.

Cultural contributions and transoceanic connections

Other intriguing archaeological finds provide clues to transoceanic connections to the Americas. A 2,000-year-old Chinese coin was found in a pottery jar by the Tlingit Indians near the village of Wrangell, Alaska. Similar Chinese coins have been found on shaman's masks. Iron chisels and tropical bamboo were unearthed from the mud that covered five longhouses at the Ozette site in Washington 500 years ago. According to Betty Meggers, ceramic pottery excavated at Valdivia, Ecuador, bears a resemblance to the 12,000-year-old "cord-marked" pottery of Jomon, Japan - the oldest known ceramics in the world. Meggers suggests this is evidence of trans-Pacific voyages from Asia to America 6,800 years ago. Divers have discovered large stones weighing up to 450 kilograms lying on the ocean floor off the coast of California. Some believe they are Chinese anchors made from stone quarried in southern China.[19]

The distribution of early plant species may also contribute to the story of early human travel. The bottle gourd, coconut, cotton, sweet potato, and corn appear to have arrived in the Americas before European contact. Although the bottle gourd and coconut may have floated from Africa and the southern Pacific to the Americas, the sweet potato and corn (maize) might have had human assistance. Maize was first domesticated in Mesoamerica and later introduced into Europe, probably by the Spanish, in the sixteenth century; however, its appearance on sculptures in India that date to the eleventh century is far more mysterious.

Migration - an option when times were bad

Clearly, there is much we do not know about the peopling of the Americas. The time and place of migration remains shrouded in obscurity; however, it is clear that climate, a responsive culture, and human resilience all played important roles. Thousands of years ago, when the place where people traditionally lived became uninhabitable, they migrated. It was that or risk death or, worse, extinction. Initially it was *Homo erectus*, and then *Homo sapiens*, who moved out of Africa. Later it was the first peoples who came to the Americas. Whether in Africa 200,000 years ago or North America 20,000 years ago, humans have used migration as a key behavioral response to a changing environment. By maintaining a close connection with the environment, people were able to survive when times were good or recognize warning signs when times were bad; when that happened, they went looking for better living conditions. They migrated. They made choices about the trade-offs between eating and freezing. They contemplated the value of tolerating some short-term discomfort for long-term gain. Failure to stay connected meant certain death.

With the onset of the relatively stable Holocene period 11,650 years ago, human populations began to control their environment. We learned how to control and exploit plants, other animals, and nature. We developed agriculture and began to settle down in one location and produce food instead of roaming in search of animals and plants to eat. Independent centers of domestication arose in southwest Asia, South America, northern China, Africa, and Mesoamerica. The next chapter delves into a few developments from this time in our human history when we began to change our environment instead of migrating out of it.

Climate and Agriculture

7

Agriculture and the Rise of Civilization

> There is no escape from agriculture except into mass starvation, and it has often led there anyway, with drought and blight.
>
> *Ronald Wright,* A Short History of Progress

Our early hunter-gatherer ancestors lived in isolated territories. One might think that, over time, their expanding populations would have exhausted resources in their territories and led them to develop agriculture, but the archaeological evidence does not support this view. Instead, hunter-gatherers practiced birth control by delaying the weaning of their children or, during the worst of times, killing the children or migrating elsewhere.

Today there is ample evidence of farming societies encroaching on the territories of hunter-gatherers and compelling them to switch to farming. There are many cases where traditional hunting and gathering grounds have been wiped out by deforestation for lumber, agriculture, or access to mineral resources. Yet these circumstances did not apply when agriculture first came into being.

Life also seems to have been a lot more leisurely for small groups of hunter-gatherers than for farmers, who had to work long hours to gain the same nutritional benefits. So making the transition to agriculture did not happen because it was less work or more fun. When resources became limited in their territories, hunter-gatherers just moved to more suitable environments, something quite possible when populations were small. There was no apparent time or energy benefit that would have encouraged them to switch to agriculture. What triggered the shift?

Archaeologists and anthropologists agree almost unanimously that it takes almost catastrophic conditions to make human beings change their ways. In 1951, anthropologist V. G. Childe came up with the "oasis theory." He suggested that a colder, dryer climate forced humans and animals to retreat to where the best sources of water remained and where hunting and gathering continued to be good. These localized groups then developed agriculture to feed the newly concentrated populations - perhaps some refugees fleeing areas of drought or destruction brought with them knowledge of upland wild cereals and stocks of their seeds. The growing population provided the labor necessary to seed, tend, irrigate, and harvest the crops. Society necessarily became stratified to organize the effort, and so began the rise of civilization.

Later researchers argued that a general increase in the human population made agriculture necessary. Likely both climate change and concentrated population increases contributed to the onset of agriculture.

Some climatic forces that affect agriculture and civilization

Like two lovers, climate and agriculture are so closely entwined that their synergy determines fertility and healthy offspring. Changes in climate influence vegetation; climate limits what can grow and dictates what will die. Climate can cause plentiful crops or catastrophic crop failures. Today, global warming is blamed for the biggest droughts in hundreds or thousands of years and the loss of entire crops. Heavy rains and floods drown crops and leave fields too wet to plough and seed. In the past, changes in climate could be just as rapid and extreme, and they played a similarly critical role in the capacity of things to grow.

Therefore, before we delve into the origin of agriculture and civilization, it is important to understand some basic climate mechanisms that have had, and continue to have, an influence on agriculture and the ability of people to live and thrive in various regions of the Earth.

The Milankovic cycles

At the heart of climate change are variations in the amount of solar radiation that reaches the Earth from the Sun relative to the amount of energy that leaves the Earth. Serbian mathematician and planetary physicist Milutin Milankovic is credited with calculating three distinct cycles in Earth's orbit around the Sun that profoundly affect variations in solar radiation.[1] The first cycle, which has a periodicity of 21,700 years, involves changes in the Earth's wobble around its axis that bring the Earth's northern hemisphere alternately closer to or farther from the Sun during the winter. When the northern hemisphere is closest to the Sun in the summer, northern seasons become much more obvious. When the northern axis is closest to the Sun in the winter, seasons are much less distinct. It is during these latter conditions that winters are warmer, more snow typically falls, and summers are cooler. These cool summers are not warm enough to melt the heavy snowfalls, so snow accumulates, forming glacial ice. Thus begins a new ice age.

The second Milankovic cycle, which has a periodicity of 98,500 years, involves the changing shape of the Earth's orbit around the Sun, from more to less elliptical. A more elliptical Earth's orbit makes northern summers cooler and northern winters warmer. Again, this reduces seasonality and encourages winter snowfall and the growth of ice sheets.

The third Milankovic cycle, which repeats every 41,000 years, relates to the tilt of the Earth as it circles around the Sun. Less tilt in Earth's axis creates less distinct seasons in the northern hemisphere. Less tilt also increases the difference in temperature between the equator and the poles, which warms the tropical oceans more and increases the intensity of the hydrological cycle in the world's oceans. This causes more El Niño–Southern Oscillation (ENSO) events.[2]

Ice sheets and greenhouse gases

During periods when northern winters accumulate heavy snowfalls and the summers are too cool to melt them, ice sheets grow. This affects the amount of sunlight reaching and warming Earth. The bigger the snow-covered area, the more sunlight is reflected back into space. This further cools the Earth's surface, acting as a major feedback that helps grow bigger ice sheets and tip the world into an ice age.

Changes in the levels of greenhouse gases like methane, carbon dioxide, and water vapor in Earth's atmosphere seem to act in unison, rising during warm periods and dropping during ice ages. The more greenhouse gases there are in the atmosphere, the more heat is trapped and the greater the rise in Earth's temperature. The fewer greenhouse gases in the atmosphere, the cooler Earth's temperature.

Note that the presence of greenhouse gases is natural and inevitable. They are essential to keep temperatures at the right level to support life on Earth.

Sea level and the productivity of the oceans

As ice sheets grow, the level of the world's oceans drops because it is from the oceans that the water comes to make the snow that falls and later becomes ice. During the last ice age, global average sea level was about 120 meters lower than it is today. Lowered sea levels exposed the productive continental shelf, where early people lived and migrated to. This is relevant to the development of agriculture. For instance, the first maize was cultivated along beach ridges and a lagoon in the Grijalva River delta on the Gulf Coast of Tabasco, Mexico, 7,000 years ago. This was 1,000 years earlier than corn was cultivated anywhere at higher elevations in Tehuacán and Oaxaca.

Since the last ice age, sea level has generally continued to rise. This is because when the Earth warms, global average sea level rises; when it cools, global average sea level falls. Sea level is currently rising at about 3.2 millimeters per year.[3] Rising sea level inundates coastal areas and river deltas where today more than 50 percent of the world's population lives within 60 kilometers of the shoreline.

The oceans are like huge heat-storage tanks. Their circulating currents distribute heat from warm to colder waters. If ocean currents are slowed or stopped, there is a big impact on the climate. Ocean currents also circulate food for marine life. Along the edges of the continents, water typically rises up from the ocean depths, bringing with it food that is crucial for coastal marine life. When this upwelling is reduced along Peru's Pacific coast, ENSO events occur; large populations of anchovies, squid, and other sea life disappear; and Peruvian coastal people go hungry. At the same time, monsoons fail in Australia, bringing droughts, while rainstorms deluge North America's west coast.

Volcanoes

Just like the 2010 volcanic eruption of Mount Eyjafjallajokul in Iceland, which interrupted global air travel and destroyed crops and fields for grazing livestock, volcanic eruptions during early-human times had short-term local effects on agriculture. About 3,000 years ago, when Mount Hekla erupted in Iceland, its impact on the environment and climate likely wiped out the majority of subsistence farmers. It has erupted numerous times in the last 1,000 years.

A volcanic eruption has less impact if people can go to farm someplace else that is not already densely populated. Even so, multiple eruptions and the occasional "super volcano" could considerably alter climate patterns and have more lasting impacts. One year of poor harvests forces people to delve into their storage of foodstuffs held for emergencies and for future sowing. A second poor or failed season brings a culture to its knees. A climate change that lasts even a few years, never mind centuries, leaves farmers and their societies devastated.

Domestication of animals

It is likely that changes in climate played an important role in the beginning of animal domestication. However, those links are uncertain and the circumstances vary with each animal domesticated. It is not likely that animal domestication arose simultaneously with plant cultivation. The first people to domesticate animals, with the exception of perhaps the dog, were nomadic pastoralists who constantly moved their temporary settlements to where their animals could graze. Droughts that dryed up lakes and reduced the amount of fodder for their animals forced the pastoralists to move. Eventually they were forced onto land inhabited by agriculturalists.

The first domesticated animal was probably the wolf, bones of which have been found in 12,000-year-old human grave sites. Genetic studies suggest there was a locus of dog domestication in East Asia beginning around 15,000 years ago. Some bold wolves scavenging the remains of kills from human hunts probably began to approach temporary human camps in search of food. Playful, tame pups drew favor and empathy from their human camp mates. Another example of a connection between domestication of tame individual animals and their subsequent evolution occurred more recently. Fur traders captured Arctic foxes, often choosing the tamest individuals. Doing so had remarkable effects. The hormone and reproduction patterns of the foxes changed and became very similar to those of domestic dogs. Their coat color and patterns also changed.[4]

Dogs were likely domesticated because they helped in hunting. It is easy to see why: just watch a border collie work a herd of sheep. The sheep bunch up without any stimulus except the presence of the dog. Other than improving the efficiency of hunting during difficult times, it is unclear how or if climate is connected with dog domestication, but dogs would also have been an accessible source of food in time of need. Even today, dog meat remains a dietary staple for some East Asians, and the survival of Inuit in the challenging climate of

Canada's Arctic has been attributed to dogs, used for hunting, transportation, and warmth.

Other animals were first domesticated by nomadic pastoralists, who set up temporary settlements with their herds and collected plants, insects, fungi, and other foodstuffs they scavenged in the course of their nomadic journeys. Later, a changing climate reduced the amount of available fodder for their animals, forcing the nomads to seek refuge in the oases occupied by established farmers. The new farm animals provided milk to drink, meat to eat, and fertilizer for the farmers' fields.

Sheep and goats were first domesticated in Mesopotamia 10,500 years ago. They were herded for their milk, meat, and wool. Mesopotamian farmers storing grains probably discovered a rapid growth in mouse and rat populations and so were keen to domesticate cats. The ancestor of the domestic cat is the wildcat, *Felis sylvestris*. The genetics of wild cats changed in the course of their development into domestic cats. Changes in their production of growth hormones made them much smaller than their ancestor.

Pigs were independently domesticated from the wild boar in southwest Asia and north China 9,500 years ago. They were introduced to Europe from China in the eighteenth and nineteenth centuries, and later hybridization of the two strains led to today's breeds of pig.[5] Pigs provided good meat, and so did chickens, which could also be plucked for their down and feathers.

The domestication of the auroch (which became the ox) was a genuine challenge because of its aggressive nature. This was accomplished over time, probably because, like humans, aurochs require a lot of water to drink and so would often encounter humans on their way to the same water holes that humans frequented. They likely were a lot less panicked about sharing this resource with humans than with prowling lions or leopards.[6] The auroch was independently domesticated in India, Africa, and Mesopotamia. Later genetic mixing resulted in the cattle breeds we see today. Their first domestication may have happened 9,500 years ago, but more likely occurred between 8,000 and 7,500 years ago. By 7,000 years ago, oxen (or cattle) were common from the Sahara/Sahel to the Middle East. Once domesticated they were used to draw ploughs and wagons, as well as providing meat, milk, and hides. Donkeys, domesticated 7,000 years ago in the Sahel of North Africa, were also used to carry loads, as were the llama and its relatives, which had the added advantage of providing wool.

The horse, camel, and elephant made transportation easier and were also used as weapons of war. The timing of equine domestication is very difficult to estimate, in part because the anatomy of domestic horses is very similar to that of wild horses, so it is hard to tell whether horses were actually domesticated or whether wild horses were just bred in captivity. Any harnesses and other tack indicating domestication would have been made of wood or hide and quickly disintegrated. Archaeologists have tentatively dated domestication to 5,600 years ago, based on corral posts found in Kazakhstan, East Asia.

Farmers cultivate plants around the world

To find our original ancestral farmers, we go to Mesopotamia, the "Fertile Crescent" in southwest Asia, which included much of what is now Iraq, as well as parts of Syria, Turkey, Iran, and Khuzestan. Known as the "cradle of civilization," the Fertile Crescent was the site of a series of empires, including Sumer, Akkad, Babylonia, and Assyria. Agriculture first started here between 13,000 and 11,500 years ago. By 10,500 years ago, many of the first plants and animals had been domesticated.[7] Farmers in the region grew emmer and einkorn wheat, barley, oats, chickpeas, lentils, and olives. Many of these cereals grew wild and fairly close together in the uplands of the Fertile Crescent.

Squash, manioc, peanuts, and cotton were first domesticated 10,000 years ago in Peru. Interestingly, wild squash was not found in Peru at the time, so archaeologists think it must have been imported by people traveling south through wild squash habitat.[8] By 9,000 years ago, farmers were cultivating chili peppers, manioc, sweet potatoes, and peanuts throughout South America. They did not start growing the potato until 5,500 years ago. Mesoamerican farmers were the first to grow corn (maize), beans, and turkey between 7,000 and 5,500 years ago. As mentioned earlier, the first recorded evidence of corn cultivation has been found along beach ridges and lagoons in the lowlands of the Grijalva River delta on the Gulf Coast of Tabasco, Mexico, dated to 7,000 years ago.[9]

In North China, farmers began growing rice, millet, and silkworms 9,500 years ago. Taro, yams, breadfruit, coconuts, and bananas were not domesticated until 6,000 years ago in Southeast Asia, whereas sorghum, African rice, peanuts, yams, and guinea fowl were all domesticated 7,000 years ago in the Sahel of North Africa, south of the Sahara Desert. It took until 5,000 years ago for African yams and oil palms to be first domesticated in tropical West Africa.

The initial appearance of agriculture in Europe is thought to have been a consequence of farmers fleeing the flooding of southwest

Table 7.1. *The timing of plant and animal domestication; points of origin of domesticated species*

Original Site of Domestication	Plants	Animals	Date of Domestication
East Asia		dog	12,000 yrs ago
SW Asia (Fertile Crescent)	emmer and einkorn wheat, barley, oats, legumes (chick peas, lentils) olives.	sheep, goats, cats	10,500 yrs ago
South America (Peru)	squash, manioc, peanuts, cotton.		10,000 yrs ago
SW Asia		pigs	9,500 yrs ago
N China	rice, millet	pigs, silkworms	9,500 yrs ago
SW Asia		cats	9,500 yrs ago
South America	chili peppers, manioc, sweet potatoes, peanuts.		9,000 yrs ago
Africa, India, Fertile Crescent (independent domestications)		cattle	9,000–7,000 yrs ago
Sahel (N Africa south of Sahara)	sorghum, African rice, peanuts, yams	guinea fowl, donkey	7,000 yrs ago
Southeast Asia	taro, yams, breadfruit, coconuts, bananas		6,000 yrs ago
East Asia (Kazakhstan)		horse	5,600 yrs ago
Mesoamerica	corn, beans, (squash ?)	turkey	7,000–5,500 yrs ago
South America	potato		5,500 yrs ago
Tropical West Africa	African yams, oil palms		5,000 yrs ago

Note: This table was originally published in Hetherington, R. and Reid, R. G. B. (2010). *The Climate Connection: Climate Change and Modern Human Evolution.* Cambridge: Cambridge University Press.

Source: Diamond, J. (1999). *Guns, Germs, and Steel: The Fates of Human Societies.* New York: Norton, with some modification from Burroughs, W. J. (2005). *Climate Change in Prehistory: The End of the Reign of Chaos.* Cambridge: Cambridge University Press and Fagan, B. (2004). *The Long Summer: How Climate Changed Civilization.* New York: Basic Books.

Asian coastal areas about 8,200 years ago, around the time of the collapse of the huge ice dam in Canada's Hudson Bay region.[10] Agriculture finally arrived in the eastern woodlands of North America after 5,000 years ago.

A closer look into why agriculture began where it did

Climate change and population growth are often thought to be the key stimulants for the adoption of agriculture and the domestication of animals; however, the climate of the Holocene (11,650 years ago to 1800 AD) was far more equable than that of the last ice age. So what was it about the shift into the Holocene that stimulated agriculture? What impacts, if any, did climate have on territories inhabited by humans and the shift to farming?

The climate 12,700 years ago in Europe and northern Eurasia was cooler than it is now. It rained less, and it became harder and harder for hunter-gatherers to find food. Around this time, giant glacial Lake Agassiz in North America flushed its near-freezing freshwater into the North Atlantic. The sudden torrent of water abruptly slowed the northerly flow of warm sea water in the Atlantic and triggered a cool, dry period, called the Younger Dryas, that lasted until 11,650 years ago. At the same time, an enormous polar desert extended across Eurasia from the edge of the European ice sheets, southward to the Mediterranean Sea and eastward into eastern China. Modern humans who had migrated into Europe became isolated from those who had migrated into the rest of Eurasia. Those in Eurasia were forced to travel into southwest Asia, India, China, and Southeast Asia to avoid the forbidding environment of the polar desert. People congregated in areas adjacent to fresh water in lakes, rivers, and springs.[11]

There were two areas where conditions were significantly better than they are today, and better than the conditions in other places at the same time, and into which people likely flocked. The first was a region in the Fertile Crescent where many archaeologists believe agriculture and civilization first began. Some tribes in the Fertile Crescent became primarily farmers, whereas others became nomadic pastoralists, especially after acquiring domestic dogs. The second region in which people gathered was an area of south Asia including the Indian subcontinent and the Indus basin where wheat, barley, jujube, sheep, and goats were domesticated beginning around 11,000 years ago.[12] As diverse human populations migrated into, and concentrated in, these two regions, human behavior began to evolve. People shared their different habits, technologies, and ways of being. They collaborated. Novel ideas emerged, including the development of agriculture and the rise of civilization.[13]

Mesopotamia, Africa, and Egypt

Abu Hureyra, one of the most famous archaeological sites in the Fertile Crescent, provides an explanation of why agriculture developed. It was settled 12,700 years ago at the beginning of the Younger Dryas cold interval. During this cold, dry period, people cultivated wild rye, but not long after the cold interval ended 11,000 years ago, they developed their own domesticated seed forms, and agriculture began. Archaeologists have found 157 species of plant seeds that were available to inhabitants of Abu Hureyra between 11,000 and 8,000 years ago. It appears that most commonly ate einkorn wheat. Archaeologists think that periods of scarcity caused by climate change, combined with the settlers' unwillingness to move from their home, where they enjoyed a good source of water and fishing, encouraged people to begin to harvest their own wheat and other crops.

Life for the inhabitants of Abu Hureyra changed when they began cultivating cereal crops. Women prepared grain for cooking, a demanding job that stressed their knees, wrists, lower back, and toes. A farmed area could support 10 to 100 times more people than hunting and gathering. This food stability, combined with the availability of cereal porridges that allowed infants to be weaned earlier, led to an increase in fertility; the population expanded. Fewer people died from illnesses, because, unlike the more susceptible hunter-gatherers, farmers and their families, over several generations, developed immunity to the diseases that farm animals carried. Although men continued to hunt, fish, and herd animals, game became more difficult to find, and it was not long before most had domesticated animals. Old female and young male sheep and goats were slaughtered for meat, although the best young males were kept to breed with the females; young females were kept for good mothering and milking qualities as well as for their wool.[14]

Throughout the cool, dry Younger Dryas, people were attracted to the continuously available water source in the not-too-distant village of Jericho Springs. Inhabitants developed agriculture and a sense of continuity. They "belonged" to a particular tract of land. Spirituality arose as loved ones buried their dead and put figurines on their shrines.

Early settlers of Euxine Lake (later to become the Black Sea) combined farming with hunting and gathering. Their diverse skill set would later prove invaluable. Although Euxine Lake lay below the level of the Mediterranean Sea, it was protected from any incursion of salt water by the Bosporus Sill. Plenty of lakeshore areas, cut by rivers and streams, made suitable farming sites. Then, 8,200 years ago, the meltwater from North America's giant Laurentide ice sheet erupted into the North Atlantic. Across the Atlantic Ocean, the level of the Mediterranean rose by 1.4 meters, breaching the Bosporus Sill. A huge influx of salt water flooded the farmlands; in what would have seemed an instant, Euxine Lake disappeared and the Black Sea appeared. The settlers of Euxine Lake spread north and east, fleeing the rising sea.[15]

Not far from the Black Sea, a diverse group of people concentrated in and around the Fertile Crescent 8,200 years ago. Monsoons swept the land. Those who had listened to the stories of their elders may have noticed that the amount of summertime sunlight (insolation) that reached them was more than usual – it was greater than it had been in nearly 100,000 years. Perhaps it was these changes in climate, combined with the different ideas and technologies that these disparate people brought with them, that allowed new agricultural techniques to be developed. Whatever the reason, agricultural production improved and new behaviors and technologies were transmitted and adopted by other settlements to help feed an increasing population. By 7,800 years ago, early Mesopotamian farmers had created irrigation canals to deliver water to crops and drainage ditches, as well as levees to control flooding. Their hierarchical social structure became more complex, and religions were institutionalized around 7,000 years ago.

Then, as the climate became dryer and dryer, the population fell. Some settlements disappeared and others grew as more and more people moved to larger towns.[16] Suddenly, 6,000 years ago, as a result of the changing orbit of the Earth around the Sun, the amount of sunshine reaching the northern hemisphere decreased. At the same time, rainfall in the Fertile Crescent dwindled. Lake levels in northern Africa began to drop precipitously, reaching their minimum levels about 2,000 years ago. Where irrigation existed and could be maintained, settlements thrived and populations increased. Greater bureaucratic control was exercised over the grain supply. In Mesopotamia, the combination of new technologies, skills, and behaviors and the stress of arid conditions led to the emergence of the first complex, urban, state-level societies.[17] Elsewhere, farmers reverted to hunting, gathering, and nomadic herding.

By 5,000 years ago, the Sumerian and Akkadian civilizations were on the rise. Organized city states competed intensely over abutting hinterland territories. Periodic droughts wiped out vulnerable settlements and intensified the growth of larger cities located in the most benign environments. An organized bureaucracy presided over irrigation, planting, harvesting, and a religious component that controlled the elements. Some leaders gathered all the cereal crops, then issued rations of grain and oil to their citizens. Standing armies, which had been in place before the Sumerian and Akkadian civilizations, became prevalent.

About 4,800 years ago, Tell Leilan on the Habur Plain, one of Sumer's largest cities, was overrun by Akkadian forces.[18] After expanding and fortifying the city, the Akkadians ruled for no more than a century, enjoying a relatively benign climate. Then, 4,100 years ago, a volcanic eruption to the north may have caused a short-term volcanic winter in the area, which, combined with a 278-year drought, caused people to flee the Habur Plain or die, leaving it deserted for three centuries. It was not until about 3,900 years ago that Tell Leilan was once again inhabited, becoming the center of the Amorite state.

In Africa, between 12,900 and 7,800 years ago, rain fell where today the Sahara Desert stretches as far as the eye can see. Vegetation grew around temporary pools, and permanent lakes filled with crocodiles and hippopotami. This was because the intertropical convergence zone (ITCZ), where trade winds converge near the equator, shifted northward, providing monsoon rains for the desert.[19] But once that lush age was over, the crocodiles and hippos disappeared, to be replaced by cattle herded on the desert grasslands. During good years, nomads found temporary forage by driving herds north to where a few small herding settlements survived around oases supplied by aquifers. When severe drought struck, herders slaughtered their prized livestock, entreating "superhuman" beings to help them deal with uncertain climate. Others led their cattle back toward the Nile, where conditions were far less difficult. Here ample food could be grown and gathered, whether it was fish from the ponds and marshes, crops from the fertile soil, or animals grazing on the plentiful pastures. Today this religious cattle-based "cult" is evident over large parts of the Sahara, perhaps spread by rapidly moving nomads in search of pasture and water between 7,300 and 5,700 years ago.[20]

It was immigration like this from the Sahara and the Nubian deserts that led to the rise of the Egyptian civilization on the Nile between 5,000 and 4,800 years ago. Population increased and new innovative agricultural techniques were developed. Agriculture was already being practiced around oases in the south of Egypt, and as those populations expanded northward, they provided the labor force needed for irrigation projects, the construction of buildings and pyramids, and the military.

As agriculture became more common, agricultural societies doubtless tended to destroy nearby habitat as a consequence of farming, unlike foragers who typically conserve wild resources for subsequent harvesting. Over time, as the returns from foraging fell, neighboring foragers were likely persuaded to change their ways and switch to agriculture. The combination of an expanding land and population base, the technical advantages of domesticated crops, the military advantages that large populations gained over smaller dispersed ones, and a conducive climate probably explains the diffusion of agriculture out of its centers of origin.[21]

When the drying climate caused a reduction in the flooding of the Nile, the centralized Egyptian government collapsed, in part because Pharaoh Pepi became senile and died. The Egyptian empire disintegrated into small settlements. By 2046 BC (about 4,056 years ago), King Mentuhotep I of Thebes reunified Egypt. He invested heavily in agriculture and centralized storage, perhaps wisely because the Nile floods continued to dwindle; however, he did not invest in new irrigation techniques. Life expectancy was low, and perhaps as a consequence, previous droughts were forgotten. At any rate, it was to the gods that they looked to maintain plenty. No contingency plans were developed.

Yet it was not long before the next major series of droughts hit. A shift in the intertropical convergence zone 3,200 years ago brought with it a drying climate.[22] Greece, Turkey, and the northern Levant became arid for several years. The Hittite empire, which had moved into the northern Levant, and the Mediterranean Mycaenian kingdom fell. Only the Egyptian kingdom remained intact.

In West Africa, a drying climate during the mid-Holocene forced people to move into previously unexplored regions, where they found new and varied food resources. As climate changed, so did the available foods. People who remained flexible and opportunistic, able to alter their dietary customs and habits, survived.[23]

The Indus-Saravati region

As noted earlier, the second region into which early people flocked to avoid the forbidding environment of the polar desert was the Indian subcontinent and the Indus basin, where conditions were significantly better than they are today. By 11,000 years ago, people had domesticated wheat, barley, jujube, sheep, and goats in small villages and pastoral camps of nomadic hunters. It was here, after the climate began to dry 8,000 years ago, that the Indus or Harappan civilization began to bloom. Lake Lunkaransar, in northwestern India, reached its height between about 7,800 and 6,500 years ago, but by 5,600 years ago it was bone dry.

During this arid period, between 6,000 and 5,300 years ago, the "early Harappan" civilization flowered. People emigrated and concentrated in the most benign areas, along rivers, where water supply was the most secure. New technologies developed as well as changes in burial practices. Within a hundred years, urban centers had developed into a society now known as the "mature Harappan." Architects designed public buildings, town planners decided where they should be built, a bureaucracy ran the affairs of government, early accountants developed and used a system of weights and measures, society became stratified, and politicians presided over it. Small villages disappeared to be replaced by larger towns. The Harappan people created a unique set of signs and symbols unlike any used before or in any adjacent region. These symbols represented a unique, probably religious, ideology. The Harappan people also developed organized trade with Mesopotamia. A rapidly drying climate drove people to explore and develop new ways of being that gave the Harappan society a unique and complex identity.

Conclusion

The cooling and drying climate of the Younger Dryas brought drought to early people. With the onset of the Holocene, they faced abrupt warming. The relatively stable climate since the beginning of the Holocene allowed people some capacity to predict seasonality and temperatures, but it was not without its own challenges. Changes in climate influenced vegetation and animal populations, affecting where people lived, what food they ate, and how they changed their behavior and technologies. Changes in the weather that persisted for one growing season typically made people alter their behavior only a little bit; longer-term catastrophic climate changes caused people to permanently change the way they did things.

Agriculture is more than just growing enough food to eat. Preparing, planting, nurturing, harvesting, and provisioning agricultural products has a big impact on how humans work, interact socially, trade, do politics, and wage warfare. Agriculture has been influenced by crisis (usually climatic), communication (of previously isolated groups of people), and collaboration that resulted in the exchange and development of agricultural techniques.

The concentration and expansion of diverse groups of people in areas where good water and food resources were available facilitated communication and collaboration. This led to the development and proliferation of new ideas and technologies. As a result, people developed agriculture and domestication, and these skills spread throughout the world. Yet subsequent climatic crises created more social instability. People concentrated in cities where the most favorable conditions existed, and the first civilizations were born. Organized states controlled irrigation, planting, harvesting, religion, trade, and standing armies.

The archaeological history of Mesopotamia and the Indus basin, outlined briefly here, provides ample evidence that climate has had an enormous impact, not only on the origin and development of agriculture and domestication, but also on the evolution of human society and civilization.

Let us now look to the other side of the world, to Mesoamerica and South America, where another civilization was simultaneously being born.

8

The Maya Civilization and Beyond

Civilization has never recognized limits to its needs.

John Perlin, A Forest Journey: The Role of Wood in the Development of Civilization

The sun has set on another day in Yucatan, Mexico, home of the great Maya civilization. According to the Maya Long Count calendar, the Maya world was created on the mythological date of August 11, 3114 BC, long before any actual Maya culture is seen in the archeological record. What began with small farming villages grew into a great civilization where religious nobility built sculptured stone monuments that registered events in the lives and history of the ruling class (see Figure 8.1). In the city of Cobá, on the northern Yucatan Peninsula, a series of white roads built between 600 and 800 AD connect a collection of buildings that stretch across an area of about 70 square kilometers. Most of the buildings were situated near the lakes of Cobá and Macanxoc. One building stands out: Nohoch-Mul, meaning "big mound." It remains today the largest structure in northern Yucatan (see Figures 8.2 and 8.3). Cobá reached its zenith during the eighth century with a population of approximately 55,000 people.

What caused the Maya civilization to flourish and then die?

Scientists studying the climate of the region have found that rainfall played a critical role in Maya civilization. It was located in Yucatan in a seasonal desert, where farmers grew agricultural crops. Maize (corn) was so important to the people that the ancient Maya would press and bind the skulls of their children so they would form a slanted cone, resembling the shape of the head of the God of Maize. The strength of their economy depended a great deal on how consistent

Figure 8.1. Sculptures from Maya city of Cobá, Yucatan Peninsula, Mexico.

Figure 8.2. Nohoch-Mul, Maya.

the precipitation was during the summer, their rainy season, because this would determine the wealth of their corn crop. Winters were typically dry.

To deal with regular variations in seasonal rainfall, the Maya developed different ways to store and accumulate water. Rock quarries and excavations were converted to water reservoirs. Buildings were often built on small hills, with canals funnelling water downhill to complex irrigation systems.

But sometimes rainfall dropped more than was normal in a typical season. Scientists have learned this by studying lake cores taken from the sediments at the bottom of Lake Chichancanab on

Figure 8.3. View of northern Yucatan from Nohoch-Mul.

the Yucatan Peninsula. The cores show three cycles of drought that affected the Maya culture.[1] The first occurred between 475 and 250 BC, when Maya civilization was just forming. At this time, agriculture was still flexible enough for people to accommodate to the stresses caused by dryer years.

The second drought happened between 125 BC and 180 AD, just as the first cities and civilizations were flowering in the lowlands. One of those cities, El Mirador, was situated in the lowlands, where water could be trapped and stored. Initially this system worked, and there was enough water for the people when the drought hit. But as more and more people congregated in the city and the drought persisted, the city grew too large and there was not enough water for everyone. Ceremonies performed by the Maya rulers failed to provide sufficient water, and the lords of El Mirador lost their spiritual credibility. The people dispersed into outlying areas, where there was still unused space to which the Maya could flee.

By 180 AD, the drought had ended and growth resumed. The Maya civilization entered a very rapid expansion. It was not long before nearly all the arable land was being cultivated. A growing population meant that even a minor drop in agricultural productivity had a big impact.

Then disaster struck. Something profound happened, and the Classic Maya civilization came to an end.

A core taken by scientists from deep in the ocean floor in the Cariaco Basin in the southern Caribbean has captured 2,000 years of climate history and helped explain the mystery.[2] In the core we can see layers of sediment that were deposited on the ocean floor in each of the past 2,000 years. These layers tell a story about how rainfall and winds in the region have changed. During wet years, a lot of sediment is deposited; during dry years, very little sediment settles on the ocean floor. The core shows that the first in a series of multiyear droughts occurred about 760 AD. After this drought, people in the region faced an extended time of little rainfall. Then, only fifty years later, another major drought hit that stretched on for nine years. In 860 AD, a third severe drought struck, lasting three years. Finally, yet another fifty years later, another drought parched the region and its people for six years. Access to groundwater steadily diminished. Insufficient rains wreaked havoc on an economy that depended heavily on agriculture. Nobility were demanding more and more crops, but their demands were becoming increasingly difficult to satisfy. And their rule was undermined when existing technology and ceremonies did not provide the much-needed water, especially in places that did not have artificial reservoirs. All this led to the great Maya demise, known as the Terminal Classic Collapse of the southern lowland. Many of the densely populated cities fell and became permanently abandoned. Some people resorted to cannibalism.[3]

The same fate befell the city of Tiwanaka, located on the Altiplano, the high mountain plain that stretches across Peru, Chile, and Bolivia. Agriculture thrived here from about 400 BC, and by 650 AD, a splendid city had developed. Irrigation systems and water storage were built. Agricultural fields were raised so that winter vegetables like potatoes were protected from frost and from being permeated with salt water. A group of outlying settlements provided the labor to maintain the land and tend the crops. Yet again, three cycles of drought between 540 and 1450 impacted the region. During the third cycle, the raised fields completely disappeared.

Why did these droughts happen, and why did they have such profound effects on agriculture and civilization?

A closer look at the climate of Mesoamerica and South America

Although agriculture developed differently in Mesoamerica and South America than it did in the Fertile Crescent, the similarity was that, in both places, climate before agriculture and civilization was different than it was after. At the beginning of the Holocene 11,650 years ago, ice that had covered the southern tip of South America began to disappear. The climate warmed and El Niño activity increased, bringing rain and warm temperatures to the coast of Peru.[4] Precipitation caused floods and mudslides. Just offshore, the cold upwelling currents that typically bring food for fish, seabirds, and mammals failed. Populations likely moved inland, where food resources were more plentiful. Those who lived in the interior benefited from improved conditions but also faced an influx of displaced coastal people.

It is around this time, 10,000 years ago, that agriculture first began in Peru with the domestication of squash, manioc, peanuts, and cotton. It is entirely possible that the crisis of a change in climate, combined with the concentration of a diverse group of people in a restricted area, increased communication between the different groups. As a result, new ideas and technologies that had perhaps only been tentatively considered up to now were implemented on a large scale to feed the expanding population.

Warm conditions lasted until about 8,200 years ago. Then the ice dam holding back the large glacial lakes in Canada's Hudson Bay collapsed, releasing an enormous amount of fresh water into the North Atlantic. This triggered a collapse of the Atlantic Ocean thermohaline circulation, which also affects the circulation of the Pacific Ocean. The result was colder and dryer conditions in coastal South America, where El Niño activity was consequently reduced. The nutrient-rich waters began to well up from the ocean depths again, improving the availability of coastal resources.

While coastal resources became more abundant, interior terrestrial resources became less plentiful, because Amazonia is generally more arid during cold intervals. Some people were motivated to move back to the coast. However, rising water levels created a refuge of sorts at Lake Titicaca in Peru.[5] People living in the interior of Peru probably hastened to the lake during this and other cold intervals to escape the drying climate and the lack of terrestrial resources. Again, a concentration of diverse peoples communicated their different ideas and ways of doing things, and new ideas and technologies were developed, tried, and implemented.

After the brief cold event 8,200 years ago, the climate warmed, stimulating an increase in El Niño activity once again, particularly by 8,000 years ago. Coastal populations were once more subjected to a drop in coastal food resources. This time, instead of moving inland, they began to trade with their neighbors and friends in interior regions. In a sense, this created a symbiotic relationship between coastal and interior towns as individual towns began to specialize in different resources.[6]

Over time, changing El Niño activity continued to play an important role in the development of Peru. The earliest dated major construction in Peru – Caral, the city of pyramids – was occupied for about 1,000 years, beginning around 4,900 years ago. The occupation coincides with a peak in El Niño activity. Caral is located just inland from coastal Peru. The implication is that thousands of people moved inland when El Niño activity peaked and relied on both fish and agriculture to survive. What is notable about this society, which was built on commerce and pleasure, was that it was maintained with a complete absence of war.[7]

The three C's and what this means

A recurring theme we see in all cases where agriculture and civilization arose, whether in Mesopotamia, Africa, Egypt, the Indus basin, or the Americas, is the recurrence of the three C's: crisis, communication, and collaboration. A climate crisis caused disparate people to move out of regions suffering deteriorating resources and conditions into productive and/or benign regions. The concentration of new groups of people within these restricted benign regions meant that they communicated more with each other about their different experiences, ideas, and ways of doing things. Communication led to collaboration, and new ideas and technologies were developed and implemented. Often a rise in population combined with continued climate deterioration meant people had to adopt new behaviors and increase resource production just to survive. Thus, the rapid rise in agriculture, civilization, and overall complexity observed from Mesopotamia to Africa, the Indus basin, and the Americas was precipitated by difficult and trying times rather than abundance and good circumstances.[8] Yet while there are similarities between these individual locations of human civilization, no two circumstances were exactly the same. Although climate crises, in particular drought, played a major role in driving the emergence of agriculture and subsequent civilization in different regions, each original contributing culture was unique, and some did not survive.

Beyond agriculture

Agriculture brought with it a rise in cultural continuity, tribal structure, and a sense of belonging to a particular tract of land. States arose as populations expanded and concentrated in productive regions. Central decision makers presided over hierarchies with full-time religious and craft specialists. To maintain control over their people, rulers needed to feed them and depended on new methods like crop rotation, the heavy plough, and irrigation to improve agricultural productivity. A surplus of labor meant there was time and support for art, architecture, philosophy, and education. A taxation system funded the military. A growing population became committed to specific tracts of land. Complex societies competed for trade networks that required a system of numbers and an alphabet that later evolved into mathematics and written language. The use of the heavy plough allowed the exploitation of clay-rich soils as easily worked sandy soils were depleted. Oxen were replaced by horses for ploughing fields. Crop rotation improved the productivity of fields. Enclosed domestic animals were fed in winter with feed gathered and stored earlier in the season. So too were humans, who ate potatoes, turnips, and dried goods. In regions that were physically or socially confined, warfare broke out, although it was not always present, as archaeological evidence in Caral indicates.

There is much we do not know about the rise of agriculture and civilization, but what we do know is that in many regions, humans developed a desire and the ability to control and manipulate the unpredictability of nature, at least to some extent. Agriculture and domestication were early manifestations of this desire. A stored winter harvest allowed people to survive unexpected short-term reductions in productivity. To manage longer-term crises, people invested in novel ways to increase production and further control and manipulate production, regardless of changes in weather or climate. People became more sedentary as a result, reluctant to move away from the agricultural infrastructure and from stored supplies and buildings in which they had invested so much time and effort. Human population expanded. Centralized states were developed, complete with specialists. Some of those civilizations thrived, others did not. The ability of people to respond quickly and adequately to climate and other changes was key to the survival of civilizations.

Through time, society continued to change, and the human desire to increase production and control our environment escalated. By about 1800, humans developed fossil fuel technology. The energy captured in coal, oil, and gas allowed humans to embark on the Industrial Revolution. More food was produced using less-productive land. We began to travel over greater distances more and more quickly. We controlled our environment and allowed a growing human population to move into less-forgiving regions of the world. Our population continue to grow.

Today, the human population is more than 7 billion people and is expected to reach 9 billion by 2045. We now not only control the environment in which we perform agriculture, but also manipulate the genetics of crops we plant to increase production. We are becoming more and more adept at utilizing and controlling Earth's resources.

Conclusion

Throughout human history, rapid changes in climate have triggered crises from which new human behaviors have emerged. These new behaviors had an enormous impact on our ability to survive and thrive. But let us be clear. Our early ancestors changed because they had to, not because they wanted to or thought it might be a good idea. They changed because their environment, or the people they loved, lived with, and socialized with, were negatively affected. There was a cost to changing, but that cost was less than the cost of inaction, which was loss of life, livelihood, cultural society, or, in the worst cases, extinction of groups, cultures, or societies. The Maya civilization is evidence of one society that tried to change in response to a rapidly changing climate, but failed to change enough. That civilization collapsed.

It is clear that the transition to agriculture and civilization have brought great benefits. Humans' capacity to innovate, produce, and proliferate is truly exceptional. Yet great costs have come with our progress – costs associated with increased inequality, violent conflict, and environmental degradation. So although we may think of our present civilization as an evolutionary adjustment or adaptation in response to climate change, it should be viewed as a suboptimal one.

In the next section, we will look at species dominance and how it is affected by climate change. We will investigate the meaning of adaptation and adaptability and how they relate to evolution, particularly ours. And we will probe the meaning of natural selection and survival of the fittest.

The Dominant Paradigm

9

Dominance Destabilized

> Once a man had begun, on account of his culture's ethos, to feel ambivalent about his own animal nature, as well as actual animals, his approach to hunting would necessarily be affected too. The respect, verging on tenderness, of the hunter-gatherer – the Bushman, say, who poured a little fresh water in the mouth of an antelope he had killed – would fall away swiftly.
>
> *Carol Lee Flinders,* Rebalancing the World

Before proceeding, I would like to emphasize the distinction between adaptation and adaptability. In the realm of human action, we commonly use the word "adaptation" to mean behavioral change. For example, we "adapt to the cold" by lighting fires and wearing additional clothes. However, in evolutionary terms, "adaptation" refers to Darwinian natural selection of genetic changes appropriate to particular conditions. In this case, "adaptation to the cold" might refer to hereditary changes in the subcutaneous tissue, hair structure, and general anatomy that would help an organism stay warm in cold conditions. An example would be the large body and thick coat of the woolly mammoth during the last ice age.

An adaptation is frequently defined as a genetically fixed trait, such as the structure of a head. Because the anatomy is genetically fixed, little modification is possible unless a genetic mutation produces a different fixed form. (This is why the Maya people resorted to pressing and binding the malleable skulls of their children: their genetics did not naturally create and form a conical skull shape despite the perceived cultural benefit.) When a genetic mutation happens, the organism cannot return to the original condition, so the evolutionary result is inflexible.

"Adaptability," in contrast to inflexible adaptation, refers to an individual organism's ability to modify or adjust its physiology or behavior to suit changing environmental conditions – for example, shivering, donning clothes, and lighting fires when the temperature drops, or pressing and binding a child's head to earn greater social respect and wealth. We might say that adaptations are gene-determined and adaptabilities are organism-determined.

Dominance versus flexibility during climate catastrophe

Throughout Earth's history, the inevitable response of species to catastrophic climate change has initially been for the number of species in the affected area to drop. Typically, the number of previously dominant species falls. In the worst cases, species face extinction because their genes cannot adapt quickly enough to respond to a rapid change in climate, and they do not have enough behavioral or physiological adaptability or flexibility to adjust.

After a dominant species becomes extinct, novel species – those plants, animals, insects, bacteria, and others that previously existed on the periphery and in limited numbers – freed from the constraints of competition with the dominant species, have a fresh new opportunity to develop and proliferate. Their time spent living on the edge of the dominant species' territory makes them nimble and flexible, ready and willing to change, and as a result of the climate catastrophe they may undergo physiological and behavioral changes that make it easier for them to adjust to the changing environment.

When the initial climate catastrophe abates and conditions stabilize, new resource and habitat opportunities become available for the previously nondominant and novel species. Before the climate catastrophe struck, scarce resources had limited the proliferation of new species; now, with the dominant species gone, conditions improving, and fewer species and individuals sharing a growing number of resources, a surplus is available.

In a stable environment, circumstances become conducive to natural selection. Newly successful species and behaviors become dominant, and adaptation, with its gradual change, prevails once more. Small, adaptive, beneficial genetic changes gradually occur over a long period of time; these changes may improve the new dominant species' capacity to compete. As the population of the newly dominant species grows, they consume more resources. Over time, the expanding population learns how to use limited resources more and more efficiently.

Meanwhile, less-successful species are once again forced into fringe environments and their numbers are limited or begin to fall.

The agriculturalists dominate during the Holocene

Based on this scenario, the relative climatic stability of the Holocene (11,650 years ago to 1800 AD) should have resulted in little change on Earth, evolutionarily speaking. This is particularly true of the last 8,000 to 10,000 years, during which climate has been remarkably stable relative to the wide-ranging climate cycles of the previous 800,000 years. During the Holocene, natural selection should have operated well, because a stable climate is conducive to slow, gradual evolutionary change. A stable unchanging climate allows dynamic stasis or stability,[1] competition, and survival of the fittest to come into play and maintains circumstances in a steady state for a prolonged period – until climate change disequilibrates things again. So during the Holocene, we would not expect humans to experience any genetic evolutionary progress. And it appears they did not.

Yet whereas adaptation through genetic evolution stalled, the human capacity for adaptability soared. The Holocene witnessed an unprecedented revolution in human culture and behavior. This revolution – or, perhaps more accurately, series of revolutions – occurred in a geological blink of an eye, far too quickly to invoke natural selection's slow and gradual evolution. So what led to such rapid human change?

Around 10,000 years ago, when *Homo sapiens* developed agriculture and domesticated animals, they created an extraordinary circumstance. Instead of being the highly mobile hunters and collectors of food they once were, they became farmers who controlled, and were tied to, a particular tract of land. Farmers converged in productive regions. Agricultural societies flourished by producing food for their growing population. By growing their own food and generating and storing a sustained surplus, agriculturalists guaranteed access to food and other resources even when times were tough. This allowed them to weather sudden environmental changes and, if they chose, to focus their efforts on activities not related to survival.

As efficiency in farming developed, more and more individuals could avoid daily toil in the fields and among the animals. These people spent their time doing other things, including communicating and collaborating with each other; some explored intellectual pursuits. New ideas were spawned and skills were developed. Religious and craft specialists appeared in populations concentrated in fertile regions. Skills and ideas were then transmitted through education and apprenticeships made possible by the growing per capita surplus, which also made possible the proliferation of hierarchical behaviors and infrastructure in the societies of the dominant agriculturalists.

As a result, the human agriculturalist population expanded further, at the expense of hunter-gatherers. The agriculturalists exercised additional control over and manipulation of the environment in the form of selective planting and irrigation techniques. Production grew and so too did the surplus. This led to a further burgeoning of the agriculturalist populations. The growing population increasingly congregated in fertile centers, where personal interaction and communication stimulated more new ideas, further benefiting these societies.

The agriculturalist population continued to increase until resources in the territories they inhabited became limited, making it more difficult to produce enough food. There was pressure to enlarge their territories and improve their techniques so they could continue to expand production. The hunter-gatherer "fringe" populations began to suffer huge losses as agriculturalists expanded farther and farther afield and as individual members left the hunter-gatherer and pastoralist groups and went to live in agriculturalist population centers. While the dominant societies boasted a variety of cultures, languages, and religions, most alternative peripheral societies dwindled, along with their unique religions, languages, and cultures.

Then, perhaps as a result of increased communication and wealth in population centers, a number of technological leaps occurred, the most well known being the Industrial Revolution, which precipitated dramatic population growth in dominant societies. Agriculturalists discovered how to use chemical fertilizers and the ancient energy stored in fossil fuels, which allowed them to generate a huge food surplus that fed an ever-expanding population. There seemed no end to the technological developments that could enhance the well-being of dominant societies and increasingly powerful corporations. Medical wonders prolonged the lives of citizens. Inexpensive, accessible methods limited human reproduction. Dominant societies became wealthier and adept at garnering resources from distant lands.

Yet conflicts, disease, mass migrations, and death continued to afflict citizens, particularly those of the less-fortunate "fringe" societies, which faced shrinking territories, environmental destruction, and diminution of resources. Available food and resources in the fringe societies declined, generating additional political and economic instability.

Climate stability ends

Sometime around the dawn of 2000, Earth reached a tipping point. *H. sapiens* began to consume more resources than Earth could replenish, and climate was increasingly affected by human activity. Despite - or, more correctly, because - of agriculture, technology, and a ballooning human population, Earth once again fell into a state of global climatic instability. Naively or not, *H. sapiens* initiated a climate change that even the most advanced technology could not buffer us from. Bizarre weather seems to be the new normal. Scientists and lay people speak of an imminent climate catastrophe the likes of which humans have never seen.

On January 13, 2010, a 7.0 magnitude earthquake hit Haiti. The devastation was immense; 1.3 million people displaced, 300,000 injured, and an estimated 316,000 dead. Two and a half months later, $12.7 billion was pledged by foreign governments, individuals, and organizations to help the people of Haiti. The outpouring of generosity was impressive.

The Haiti earthquake was a natural disaster, not a climate change event. But it was followed by what is becoming an all-too-frequent event in Haiti - heavy rains causing flooding. Hundreds of thousands of people were living in temporary shelters amid the stench of human waste. Diarrhea, abdominal pain, and fever were soon widespread, as was cholera. Flooding and extreme rainfall exacerbated an already tenuous situation.

Six and a half months later, devastating floods hit Pakistan; 20 million people were affected. Floods washed away crops, submerged villages, and killed an estimated 1,700 people. Then on March 11, 2011, a 9.0-magnitude earthquake, the fourth-largest in the world since 1900 and the most powerful ever recorded in Japan's history, struck that country's northeast coast. It triggered a devastating tsunami that killed thousands of people, destroyed buildings, roads, and power lines, and severely damaged the Fukushima Dai-ichi nuclear power plant, resulting in serious radioactive leaks and contamination.

Although Japan's Tohoku earthquake was not triggered by climate change, it led to an environmental crisis and, along with a growing list of global climate and environmental crises, had a profound effect on the local and global economies and markets that were already unsteady from record high levels of indebtedness.

Climate scientists predict there will be more and more extreme climate events, ushering in widespread fear and trepidation of what is to come. Governments laden with debt, depressed economies, previous commitments, a short-term focus, and weak voter support are reluctant to act.

On July 20, 2011, the United Nations declared a famine in Somalia as East Africa was hit by the worst drought in half a century. Drought combined with conflict, poverty, limited access to aid, and rising food prices put millions of people at risk. Half of the Somali population – 3.7 million people – were in crisis, with hundreds of thousands seeking refuge in camps in Kenya and Ethiopia. According to one report, 30 percent of children were acutely malnourished. Aid agencies, governments, and local residents in refuge areas came to the rescue. But the food crisis spread to Ethiopia and Kenya.

Will the world give enough, soon enough? Will a crisis- and donation-fatigued world be there for the next catastrophic event? If we are less generous, what will that mean for humanity? Will the next catastrophe further destabilize an already fragile global economy? Will conflict escalate?

Although climate is frequently blamed for triggering conflict, evidence of its responsibility is sketchy. But in August 2011, scientists Solomon Hsian, Kyle Meng, and Mark Cane published research that showed examples of climate change directly linked to civil conflict.[2] Tropical countries are twice as likely to suffer civil unrest during El Niño years, which occur every two to seven years, compared with La Niña years. Apparently when El Niño warms ocean waters in the eastern Pacific, it brings scorching heat or dry winds to Africa, Australia, and Southeast Asia, playing havoc with the goodwill of residents in those countries and regions. Hsian and colleagues say that El Niño has played a role in 21 percent of civil conflict since 1950.

What does the future bring?

If dire climate predictions are correct, and if history is destined to repeat itself, then evolutionary law will prevail. As in the past, when the next climate catastrophe strikes, those species with limited flexibility will suffer the greatest impact. For humans, it will be the elderly; the very young; those dependent on urban infrastructure to keep us warm or cool, fed, watered, safe, and transported; those already living in marginal regions and on overextended resources; and those who refuse to accept change. In other words, most of us.

Human societies and organizations, particularly dominant ones, have become so sheltered from environmental change that our capacity to adjust to nature's minor nuances, let alone major fluctuations, has become severely hampered. We are dependent on fossil fuels to maintain our homes at the perfect temperature, to allow us to work and get to work, to stimulate our economies, and to facilitate consumption that fuels our economies. However, fossil fuels will be of little use when they have become prohibitively expensive and our infrastructures are destabilized by climate change.

Of all the *Homo* species, only a single one remains – *Homo sapiens*. We remain a competitive species. Instead of competing with other *Homo* species, we compete mercilessly with Earth's many and varied organisms and with less-dominant human cultures. Competition, a cornerstone of Darwin's theory of survival of the fittest, operates well when the environment is stable. It facilitates slow, gradual change and maintains the status quo of the prevailing dominant species. But as the trail of now-extinct dominant species tells us, gradual genetic change does not happen fast enough to adjust to rapid environmental change. It becomes essentially inoperative during catastrophic change.

Earth's impending climate change will be relatively rapid and significant. Do not wait in eager anticipation for natural selection to bail us out. Do not expect dominant human societies and organizations, accustomed to competition, to successfully compete with Earth, control her wanton ways, and haul us kicking and screaming from her turbulent, warm, and rising seas. Humans simply do not possess that much power – here recall the 2011 earthquake and devastating tsunami in Japan. And although dominant societies and organizations have been generous to those in need in the past, their resources are limited and our habit is competition. Today's dominance has been built by outcompeting "fringe" societies and all other species on the planet over the last 10,000 years. If the behavior of dominant societies continues to be driven by the paradigm that competition proves supreme fitness, we will indeed be living in a dangerous climate, because such an attitude is no longer relevant on a destabilized Earth.

Let us now delve deeper into that dominant paradigm of natural selection and survival of the fittest, its wonders and its follies.

10

Fitness Folly

> The spectators hate to be told that the Emperor is naked, because they knew it all along, and it is less humbling to continue the mass pretence than to come out and admit self-delusion.
>
> *Robert G. B. Reid,* Biological Emergences: Evolution by Natural Experiment

Humans have made such great strides since our origin some 200,000 years ago. We have survived all manner of obstacles when all our fellow species did not whether as a consequence of direct competition with *Homo sapiens* or otherwise. Humans have learned to control nature. We grow crops, influence genetic diversity, contain rivers, irrigate deserts, and build power-generating dams, nuclear reactors, and coal-burning hydroelectricity stations. We can heal the sick, educate the young, pray to our varied gods, fly into outer space, and eat genetically modified corn or rice grown in the desert. We can talk to our loved ones halfway around the world, send text messages and e-mails through the ether, watch the news as it happens, and listen to the symphony. We create beautiful gardens, sculpt, paint, write poetry, create music, enjoy movies, and experience wonderful meals. We can talk, sing, dance, run, cycle, swim, and fly. We develop businesses to help us achieve and consume anything our heart desires. There is little that we cannot do.

This modern society we have created has been profoundly influenced by Charles Darwin's (1809–82) theory of "natural selection," with its associated concepts of "competition" and "survival of the fittest," and by Adam Smith's (1723–90) theory of "the invisible hand." Darwin's and Smith's impacts have been biblical in proportion; their ideas pervade every aspect of our modern lives. It is through competition that the strongest survive, breed, and procreate. Self-interest and unfettered competition form the basis of the free-market system, which, according to Adam Smith, must be left free of interference, moderated only by the "invisible hand," which dictates that consumers buy and producers sell in order to increase the general economic well-being of everyone and generate the most effective and efficient use of a nation's resources. The market freely decides what is best for society and, in turn, society reaps the greatest reward. In theory, the result is a society composed of increasingly stronger, healthier individuals and businesses.

In reality, there is a price to be paid: the weak are cast aside as unfit; the environment and other species are seen as disposable resources to be used to generate a stronger economy; unprecedented waste pollutes the planet; and companies grow complacent with success and lose the capacity to innovate or adjust to change.

Interestingly, despite the fact that Charles Darwin's theory of evolution (which suggested that humans and other species developed over a very long period) contradicted the creationist beliefs of Christianity (which stated that humans, namely Adam and Eve, appeared in fully modern form in 4004 BC), Darwin never completely renounced his religious beliefs, although he became progressively more agnostic as he grew older. Nor did Adam Smith leave the economically unfortunate entirely to the vagaries of the marketplace, to survive or not as competition deemed fit; he donated large sums to charity. Smith's generous behavior toward the economically and socially bereft suggests he was not entirely committed to his theory that there should be no interference in the free-market system. Perhaps he suffered from second thoughts, guilt, pity, sorrow, generosity, sadness, compassion, or any number of feelings that inhabit a place distant from the cold-blooded world of competition.

Neither Smith nor Darwin took an exclusionist stance. They saw there was room for contradictions and exceptions. Darwin himself intimated that it was sometimes possible for various and sundry varieties, deemed "monsters," to occur and survive despite, and contrary to, the theory of natural selection, "their preservation [dependent] on unusually favourable conditions."[1] Such monsters frequently lived on the edge of the "normal" world. Richard Goldschmidt later called these mutant organisms that were produced in a single generation "hopeful monsters."

These inconsistencies do not mean that Darwin's or Smith's theories lack value and should be dismissed in their entirety. Both have provided important contributions to our understanding of the world.

Darwin's interpretation of the history of life has proved irrefutable, but there are doubts about whether his theory of natural selection is entirely correct. Smith's economic ideas and Darwin's natural-selection theory are not facts and will benefit from inquiry. And inquire we must, for it is increasingly evident that a worldview focused solely on one perspective – natural selection through survival of the fittest and competition – may be flawed because it does not explain all that we now witness.

With the demise of all other *Homo* species, humans compete vigorously among themselves and against every other species on the planet, wantonly wasteful, seemingly oblivious to our mutual dependence. Our competitive urges see no limits. We pit one society against another, one religious doctrine against another, one economic or social class against another, and one race, gender, or generation against another, believing with religious determination that whoever survives shall be deemed most fit, most capable of future dominance, most wise. Beneath our modern *H. sapiens* surface, we cling to the belief that those of us who have survived will continue to survive precisely because we are most fit. We also trust that those who control nature cannot be controlled by nature.

Thomas Robert Malthus (1766–1834), renowned for his *Essay on the Principle of Population* (1798), influenced the British government to change the poor laws to reinforce his idea that the poor were unfit and should be left without government help. Both Darwin and Alfred Russel Wallace (1823–1913), who propounded the theory of evolution by natural selection, were influenced by Malthus. So, in a sense, the idea of evolutionary fitness came out of the idea of economic fitness. Those who are unfit should die and will die, but those who are most competitively fit will continue to survive. Those who are most numerous, strongest, least able or willing to control their sexual drive, most uncaring about the welfare of any other than their genetic offspring will survive, procreate, and control Earth and all its species. In other words, they will be most successful.

The risk, of course, is that we may be wrong. Even the strongest individuals among us do die. Some never leave offspring. Even the most dominant species in Earth's history have suffered extinction. So too have previously dominant societies. Dominance does not and has not guaranteed indefinite survival. When Darwin suggested that natural selection would allow the fittest to survive and dominate, he could not have imagined how dominant and controlling we would become. Although natural selection continues to work its gradual change, humans have altered the playing field on which natural selection operates. Now we need rapid flexible change so that we can adjust to the rapid climatic changes that humans, in all likelihood, have unleashed. As the current dominant species on Earth, we have overlooked the fact that if we mess up the environment, we open the way for a more flexible species to replace us.

We may not wish to face our mortality, but it faces us. Our emphasis on competition and survival of the fittest, to the exclusion of all other considerations, engenders irresponsible behavior that is causing harm to other humans, our children, other species, the environment, and, yes, even our economy – all of which we rely on for survival. We mistakenly believe that because we have reached a position of dominance and control, we can remain there.

There is also a danger that the dominant paradigm allows us to believe we are exempt from natural life forces and beyond harm's way, that we actually *are* in control, that the environment is ours to control, and that we can blithely go about our business without worry that the environment may strike back. In our minds we have made "Mother" Earth easy to exploit and subjugate. "Giving nature a traditional feminine image is reassuring to us for 'surely a mother will always be loving toward us, continue to feed us, clothe us, and carry away our wastes, and never kill us no matter how much toxic waste we put into the soil or CFCs into the ozone.'"[2] We are naïve. We perceive the feminine to be so good and so nice – the "too-good mother."[3] We refuse to see her dark side, death, which balances and provides for life.

We lull ourselves into believing that we will survive in any event, under any circumstances, because we are "most fit," and our future offspring will not be hurt by our actions. Because we have survived, they will survive.

Yet we are already damaging our planet, placing our children and grandchildren in a precarious state. If the entire purpose of our species is to create the conditions that will allow the greatest chance of survival for our offspring – whether in the form of genes, individuals, or the species in its entirety – we would not, could not have forced a climate change that will continue to cause species extinctions, crop failures, water shortages, extreme events, and rising temperatures that push our capacity to survive into the "danger zone."

Natural selection operates best in a stable environment. Slow change reinforces the dynamic stasis of the dominant species and resists large-scale "monstrous" change. Under conditions of sudden environmental change, when rapid adjustments are required, the slow processes of survival of the fittest and competition may fail to generate the kind of results needed. We need alternative processes for intervals of rapid change. We need to rethink our long-held theories and paradigms that drive our systems, our behavior, and our thought processes.

Our exclusionary, competitive, survival-against-all-odds attitude is based on the belief that our burgeoning human presence represents expanding wisdom and fitness. That belief feeds a ubiquitous denial of death. We cannot accept that we will die, that our loved ones, our religious ideologies, our customs, societies, laws, political systems, world views, lifeways will die – or that our world as we know it today may end. Our survival-against-all-odds attitude represses fear, particularly fear of death.

But as we repress fear of death, we replace it with fear of a broadened worldview. This fear is perpetuated by politicians, border-control guards, terrorists, soldiers, parents, teachers, and religious leaders and is now inoculating our children. Fear strengthens exclusionary ideas about who is with us and who is against us. Fear heightens awareness of difference, whether in skin color, language, class, religion, or political persuasion. Fear takes us to the abyss of knowing and keeps us there, toes curled on the edge of the unknown, eyes tightly shut, curiosity curtailed, locked up in alarmed cars and gated communities with tapped phone lines and videotaped public squares.

If our species is to survive and our current societies are to be maintained, we need to step out of imposed fear, face our own mortality, and celebrate *H. sapiens*' variability. Doing so will breed compassion and the capacity to change, and it is entirely possible that we will be turning to variability when we have depleted all the resources within our current paradigm.

Darwinian natural selection will not save humanity from climate change. Slow-changing genetically fixed traits allow for little modification except through gene mutation to a differently fixed form. Once that happens, we cannot return to our original condition; we are adaptationally inflexible. Darwin's natural selection does not explain life's origin but, rather, its restriction, the reduction in the variety of form which, over time, becomes perilously dependent on an unchanging environment. When that environment changes, the adaptationally inflexible species faces potential destruction.

We need to look elsewhere for opportunities for change. What about those hopeful monsters, which were dismissed because they did not seem well-suited to the prevailing conditions of life? Perhaps they have the ability to thrive in a new environment. Perhaps their degree of adaptability or some amount of physiological and behavioral modification could allow them to adjust to changing conditions. Let us look at some less "fit" survivors and meet some hopeful monsters.

11

Darwin the Selector

> No one theory is ever entirely correct, no one human wholly perfect, and no one species forever dominant.

Only the fittest survive in the military. Physical imperfections, although conceivably unavoidable, are frequently unacceptable. When his superiors discovered that my father, at the age of thirty, had developed multiple sclerosis, they informed him that after one last promotion he would not be eligible for further promotions and offered him a single posting to anywhere in Canada – should he decide to stay on. With a family of four, soon to be five, to support and only one year of university under his belt, my father had few options, so he took the military's offer and we moved to Vancouver, his hometown.

My father dragged himself up the stairs of Canadian Forces Base Jericho headquarters to his second-floor office five days a week and frequently on Saturdays. Relocating his office to the first floor was not an option, and the building did not have an elevator. When the day arrived that my father had to crawl up the remaining few stairs and down the hall to his office, retirement was unavoidable. But fit or not, and despite a brutally difficult divorce from my mother, he had managed to work long enough to see all their children through high school.

My father exhorted me to attend university. I knew from an early age that although I was expected to attend, I would have to pay my own expenses. And although I knew that I came from genes that were not quite "fit," I persevered. Holding down two and often three jobs at a time, I obtained first a bachelor's degree in arts and then a master's degree in business administration. After a tumultuous first marriage left me a single mother with two young children, I met a wonderful man who did not buy into the Darwinistic clichés. We married and raised our two sons, and with his encouragement I returned to university to obtain a PhD and ponder the questions that had long been puzzling me: "What makes us human?" and "How did we come to be who and where we are today?"

Throughout my university career, Darwin's theories of natural selection and survival of the fittest were called on to explain just about everything about life itself. Whether in business, economics, science, or anthropology, Darwin's theories enlightened the confused and uninitiated. Survival of the fittest was invoked to justify the behavior of genes, cells, and organisms; *Homo sapiens'* ascent to the "top" of the animal kingdom; or the behavior of businesses, chief executive officers, presidents of large multinational corporations, and the market economy in general. Natural selection and competition were cited as explanations for human behavior, whether in terms of consumers, reproductive organisms, or conglomerations of genetic material. The concepts of competition and survival of the dominant society and organizations were used to rationalize difficult decisions and justify "unavoidable" atrocities on the road to progressive change, whether it was mining diamonds in Africa, controlling production of oil in Iraq, or developing tracts of wilderness.

The idea of survival of the fittest applied equally to the educational system, both in university and in the elementary and high schools that our sons attended. Competition permeated the modern world. Bigger was always better. At five foot three inches tall, I did not see the fairness of this. But as a late dear friend said, "Life's not fair," and I competed to receive the highest marks in order to be deemed more "successful." Perhaps it is understandable that my obsession with Darwinism did not dim, even after I obtained an interdisciplinary PhD.

But after a sojourn in the fast lane of the business world and a swim in the depths of the primordial soup of biological and geological life, I became aware that something was amiss. Climatologist Andrew Weaver saw the evidence when he and researchers at the University of Victoria Climate Modelling Lab predicted imminent rapid global climate change linked to the increase in the amount of greenhouse gases observed in Earth's atmosphere since the Industrial Revolution. Andrew wondered what this meant for humanity and, in particular, for the next generation, which included his two young children. Using a model that simulated Earth's climate over the past 135,000 years, Andrew and I hoped to gain a better understanding of how ice ages and warm climates might have affected humans in the past. If we could get a clearer understanding of our past behavior, maybe we would be better able to respond to future climate change, particularly future greenhouse-gas-induced warming. With an ice-sheet model contributed by Sean Marshall at the University of Calgary and other scientists from around the world, Mike Eby, Ed Wiebe, Andrew, and I spent more than a year preparing and running the climate simulations. But we needed an evolutionary biologist to help us understand how the environment influences the human capacity to adjust.

Evolutionary biologist Robert Reid speculated in his first book, *Evolutionary Theory: The Unfinished Synthesis* (1985), and later in *Biological Emergences: Evolution by Natural Experiment* (2007), that Charles Darwin did not have it quite right. Not only did Darwin's *Origin of Species* (1859) not explain the true origin of species; it also suggested that evolution was restricted to slow and gradual genetic change. According to Darwin, the "struggle for existence [leads] to the preservation of profitable deviations of structure or instinct." Natural selection acts on an "interminable number of intermediate forms" that accumulate "for the good of the individual possessor ... linking together all the species in each group by gradations."[1]

However, Reid observed that the process of natural selection actually generates stasis. Instead of stimulating the origin of new species, it limits the number of species to the "fittest" and restricts the emergence of novelties. Geological and biological evidence of the Earth's history indicates repeated elimination of previously dominant species and rapid diversification of new organisms on the heels of catastrophic climate change, with no evidence of the intermediate, linking fossil species that Darwin describes. In essence, during periods of environmental stability, natural selection actually maintains the dominant species as they drift slowly along an evolutionary path and restricts the production of new species. Rapid environmental or catastrophic change disrupts the equilibrium maintained by natural selection by removing previously dominant species and clearing new environments for habitation by previously less-dominant and novel species.

Niles Eldredge and Stephen Jay Gould have also shown that the histories of species were not marked by gradual change but by long periods of stasis.[2] Most change occurred during short periods, and it was during these periods of rapid change that new species proliferated.

All organisms, including humans, have the potential to experience this rapid proliferation of new species. Earlier in our history, there were many *Homo* species, including the Neanderthals and *H. floresiensis* described in "The Evolution of the *Homo* Species." These cousin species likely arose when the environment was changing rapidly; however, they did not survive. So, unlike virtually every other group of organisms on the planet today, humans are alone – the sole remaining *Homo* species. And although the history of *H. sapiens* has spanned 200,000 years, it has only been within the last 11,650 years that *H. sapiens* have dominated the Earth. During this stable period of the Holocene, natural selection and survival of the fittest promoted the dominance of *H. sapiens* and limited the proliferation of novel *Homo* species.

If Darwin's theories explain not the beginning of life but its restriction and the limitation of novel life, what becomes of a society and humanity based on this construct? If Gould, Eldredge, and Reid are correct in their claim that Darwin did not have it quite right, our conception of the world and our understanding of life, including our future economic development, may be based on a partially flawed or at least an incomplete paradigm. What happens to the dominant *H. sapiens* species when environmental stability shifts to instability, especially when this species is so confident of its ability to control that environment?

More worrisome still, the theories of natural selection and survival of the fittest have become so ingrained in our understanding of life and how it works that they are used as explanations and justifications for much of our behavior, excluding all other possible accounts and interpretations – curtailing innovative thinking. We use this analogy from the natural world to shape our societies and our economies. However, as sentient beings, we have the capacity for foresight and the ability to set priorities beyond the immediate gratification of biological drives. Will we allow ourselves to look beyond natural selection, survival of the fittest, and competition to find new ways of living that will allow us to adjust to impending change?

Unlike other fields of science, the discipline of biology has limited itself to a single theoretical paradigm; however, no one theory is ever entirely correct, no one human wholly perfect, and no one species forever dominant. Darwin the Selector, the simple substitute for God the Creator, provides only a single perspective on the complexity of life. And yet we continue to apply this theoretical paradigm with uncompromising insistence in all of life's situations. As the story of Woody Guthrie in the next chapter shows, natural selection and survival of the fittest do not explain our individual uniqueness, which reaches far beyond the genetic level.

12

Hunting Down Woody

The DNA keys of the genetic keyboard are necessary if the music is to be played, but they are neither the player nor the score.

Robert G. B. Reid, Biological Emergences: Evolution by Natural Experiment

In my early teenage years, I used to slip down to my older brothers' room in the basement when they were not home and listen to their record collection, especially the Beatles' *Abbey Road*, Janis Joplin's *Pearl*, and, at Christmas, a record by Herb Alpert and his Tijuana Brass. I would search through the stack of LPs and belt out Woody Guthrie's "This Land Is Our Land" and "The Farm-Labor Train," and his son Arlo's rendition of Steve Goodman's "City of New Orleans."

Married three times, Woody Guthrie and his wives had eight children. It was during his third marriage that his increasingly unpredictable behavior was finally diagnosed as Huntington's disease – the same degenerative disease that had led to his mother's institutionalization thirty years previously. He died on October 3, 1967.

Woody developed the disease at a younger age than his mother did. This type of pathological condition, in which each successive generation develops the disease earlier than the previous one, is called anticipation. Huntington's disease is caused by the amplification, or the increase in number, of a particular codon, a unit of genetic coding (a codon consists of three nucleotides that form into a single amino acid. Amino acids are the building blocks of DNA). Amplification of a particular codon occurs when eggs and sperm are being produced and is caused by a slippage during replication. With each succeeding generation, more of the codon can be replicated. When more than fifty copies of the codon have been generated, the disease becomes apparent. It is hard to predict how long it will take to make fifty copies of a damaged codon because in some replications there would not be any slippage in that part of the codon. The original codon slippage is a chance event; it happens by accident. But once it happens, we can anticipate that the disease will keep getting worse.

It is possible to imagine conditions where this type of genetic amplification would be beneficial. For example, increased codon repetition provides dog breeders with enough variation to select facial and limb-length features. Repetitions proceed until there is something obvious and new for the dog breeders or, in the case of other canids and a variety of other mammals (including otters, walruses, rabbits, bats, and humans), until it provides an advantageous condition that helps the mechanism spread.[1] Under these circumstances, according to Darwin's theory of natural selection, each successive generation should be genetically "fitter" than the previous one and more competitively advantaged. Darwin assumed that the fitter an organism, the better it is at adapting to its environment and the greater its ability to pass on its genes to the next generation.

More recently, neo-Darwinists redefined natural selection as "differential survival and reproduction." For neo-Darwinists, "fitness" is expressed as the percentage of distribution of particular genetic variants. The most numerous variant is considered the fittest. The greater the number of genes with high fitness, the greater their contribution to the *relative* fitness of the whole organism, and thus the greater the likelihood of its survival and reproduction, as well as the survival and reproduction of its "superior" genes.

However, Woody would probably not have agreed that the successive genetic alterations associated with Huntington's disease led to his "increased fitness" or bestowed on him a successful competitive advantage. Instead, Woody might have suggested that these genetic alterations were counterproductive. According to the theory of natural selection, negative genetic alterations in the codons, such as those seen in Huntington's disease, should be removed at each stage because they are "negatively advantageous" – harmful rather than helpful. After the first accidental occurrence, each successive codon amplification is totally predictable, yet the "unfit" genes continue to be produced. This is inconsistent with the premises of natural selection; the runaway genetic process is detrimental to the host, but it continues unmitigated.

According to the theory of natural selection, "sexual selection" causes mates to choose each other based on the potential for reproductive success. This presented Darwin with a dilemma, because he saw that choosing a mate on the basis of sexual attractiveness might run counter to final reproductive success. For example, in the case of birds, sexually attractive, gaudy feathers might expose their owners to predation.[2] He saw sexual selection operating at the individual level. Neo-Darwinists, such as John Maynard Smith (1920–2004), George C. Williams (b. 1926), and Richard Dawkins (b. 1941), reduced it to the gene level, recognizing that genes had the capacity to be "selfish." These "selfish genes" strive for reproductive success, cooperate with other genes only for the purpose of self-replication, and use their host organisms merely as mechanisms of transportation. Populations of genes compete against other populations; the fittest gene populations have the greatest distribution and are most successful at appearing in subsequent generations.

But in the Guthries' case, the logic of natural selection and sexual selection does not make a lot of sense. The genes are continuing to mutate in a predictable way, their numbers and distribution are expanding, but that expansion will eventually result in their hosts' demise. Further, Woody had not one but three wives, and he had children with all three of them. Neo-Darwinists might consider all three of Woody's wives somehow unfit because they chose to have children with him, even though their genes will experience diminished distribution as their hosts die at progressively younger ages.[3]

On the other hand, perhaps Guthrie's wives were attracted to him less for his genes and more for his personality. An unfathomable uniqueness exists for these and all human individuals, which belies genes. The integration of the human body, mind, and soul creates a whole greater than the sum of the parts or the genes. In the case of the Guthries, Woody, Arlo, and their families have succeeded in generating a greater global human community through their songs and their convictions.

So if there is something more to the story than natural selection and survival of the fittest, what is it? Has anyone ever talked about other evolutionary options? And if they have, why haven't we heard much about them?

A few brave souls *have* dared to question the ubiquitous application of natural selection and survival of the fittest. Paul Kammerer was one. Whether he was right or not, his story illustrates how difficult it is for those with ideas that diverge from the dominant paradigm to gain credence.

13

Kammerer's Suicide

> A new scientific truth does not triumph by convincing its opponents and making them see the light, but rather because its opponents eventually die, and a new generation grows up that is familiar with it.
>
> *Max Planck,* Scientific Autobiography

According to Darwin's theory, natural selection is not tolerant of difference or change unless it is either "neutral" or of immediate benefit to the species affected. In the same way, the scientific establishment is not always kind to the progenitors of novel ideas that conflict with the generally accepted paradigm.

In 1880, a son was born to Sofie and Karl Kammerer in Vienna, Austria. Paul Kammerer initially studied music but later switched to zoology at the University of Vienna, obtaining his PhD at the age of twenty-four. He published his findings on the care and welfare of captive amphibians, then turned his attention to an aspect of evolutionary theory that has subsequently been rejected by neo-Darwinists: Jean-Baptiste Lamarck's theory of the "inheritance of acquired characteristics." Neither Kammerer nor his chosen research field fit well within the research environment of his time, which may explain the fate that befell him.

As was the case for most Jewish biologists at the time, Kammerer had difficulty finding research work. Aware of these difficulties, Hans Przibram acquired the "Vivarium," a large building in Vienna that had previously been used as an indoor zoo, and converted it into the Institute for Experimental Biology. Przibram hired Kammerer as planner and curator. The newly renovated institute contained large "environmental chambers," an experimental feature that housed the amphibians and reptiles that were on display and used in experiments.

Scientists could control the light, temperature, and humidity of these chambers – a striking contrast to the cramped quarters available in other scientific laboratories.

Free to follow his biological research interests, Kammerer studied salamanders. In his experiments, he was hoping to observe evidence of the fourth law that had been put forward by Jean-Baptiste Lamarck (1744–1829) in his *Histoire naturelle des animaux sans vertebres* (1815). This fourth law suggested that "all which has been acquired, laid down, or changed in the organization of individuals in the course of their life is conserved by generation and transmitted to the new individuals which proceed from those which have undergone those changes." In other words, the changes that an organism's structure undergoes as it responds to its environment through life will be passed on to its offspring.

While doing his research at the Institute for Experimental Biology, Kammerer discovered that exposure to light made salamanders lighter in color. Further, they passed this characteristic on to their offspring. This finding has never been seriously disputed, and it meets the definition of "inheritance of acquired characteristics."

Kammerer also studied sea squirts and claimed that if their siphons were cut off during development, the resulting stumpy siphons appeared in the next generation. However, no scientist has achieved this same result in spite of several serious attempts to replicate it.

Kammerer's cause célèbre arose from his work on the midwife toad. This species of toad typically mates on dry land, but Kammerer was able to make them mate in water. As a result, the male midwife toads developed swollen, coarse-skinned, dark-pigmented nuptial pads on their toes, which allowed the males to hold on to the slippery females. Although the toads passed the characteristic on to their male offspring, Kammerer did not regard this strictly as the inheritance of an environmentally induced character. Rather, he believed his experiment had stimulated an ancestral feature that was found in many other toads but had been repressed in the midwife toad. However, the press trumpeted it as an example of Lamarckian inheritance.

Other biologists, especially in England, were unimpressed. The English geneticist William Bateson regarded the research as fraudulent.

By the early twentieth century, neo-Lamarckism dominated evolutionary theory in the United States. Neo-Lamarckists believed that the environment imposes changes directly on the organism and that this change is evolutionary, passed from one generation to the next.

Neo-Lamarckists also believed that the environment might have negative influences and that natural selection was thus an arbiter of failure and success. However, European scientists tended to be Darwinists, and the idea that acquired characteristics were inherited was not supported by such leading researchers as August Weismann and Ernst Haeckel of Germany or by the British scientists Thomas Henry Huxley, Francis Galton, and William Bateson, who had all received the Darwin Medal (presented to scientists for distinctive biological research in the field in which Charles Darwin worked).

Kammerer's research was halted when the First World War broke out and the institute was virtually shut down. Kammerer was assigned to the censor's office in Vienna. He was unable to maintain a breeding population of midwife toads, so all that remained for the curious to examine was one preserved specimen and some slides with sections of the nuptial pads.

After the war, with the institute left destitute, Kammerer resorted to lecturing to survive. He was received with hostility by British biologists, especially William Bateson, who even refused to examine the material that Kammerer had brought with him. Nevertheless, the Austrian was popular with the undergraduates.

When American biologist G. K. Noble, already in the anti-Kammerer camp, was allowed to investigate the final specimen, he discovered that the nuptial pad, which had supposedly been developed for mating in water, had no coarse-skinned, dark-pigmented texture and that the swelling was a result of injected Indian ink. Przibram was present and confirmed Noble's findings. The two men wrote a report for the scientific journal *Nature* in 1926.

Noble doubted that the midwife toads in Kammerer's lab had ever developed the coarse-skinned nuptial pads, and he believed that Kammerer had cheated on all his work. Przibram, however, had been present during the active research and when the slides of the pads were prepared, and he was able to confirm the authenticity of Kammerer's findings. No one knew who had doctored the surviving specimen, nor why it had been tampered with. Over the previous decade, a consensus had been building, driven mainly by Bateson, with support from Noble and others, that Kammerer faked his research results. That consensus prevailed.

Kammerer received an offer to head a new department at the Pavlov Institute in Russia, where his socialist and neo-Lamarckian ideas were held in high esteem, but the desolate scientist shot himself. His suicide was interpreted as a confession of guilt, and today reference to Kammerer, Lamarck, or inheritance of acquired characteristics elicits ridicule. Even so, the belief persists that, on occasion, changes imposed on organisms by the environment become heritable.[1] If true, it would mean that the slow, gradual genetic change associated with natural selection and survival of the fittest is not the only mechanism generating "evolutionary" change. It would also mean that "hopeful monsters," those mutant organisms produced in a single generation, might exist – and even though they are not dominant, they might possess the flexibility and nimbleness necessary to survive in a rapidly changing environment. In the next chapter, we look at the origin of the hopeful-monster concept and consider its viability.

14

Giants and Pygmies

> Under domestication monstrosities sometimes occur which resemble normal structures in widely different animals.
>
> *Charles Darwin,* On the Origin of Species

Like Paul Kammerer, Richard Benedict Goldschmidt was discredited because his novel ideas ran counter to the received wisdom of the scientific establishment. Born in Frankfurt in 1878, Goldschmidt studied morphology and embryology and later became director of genetics at the Kaiser Wilhelm Institute for Biology in Berlin. Twice his career was interrupted by war and politics. He was placed in an internment camp in the United States as he traveled from Japan back to Germany during the First World War, and after resuming his position at the Kaiser Wilhelm Institute, Nazi persecution forced his departure. He resurfaced at the University of California Berkeley in 1936. However, his unorthodox views cast him in direct opposition to neo-Darwinists.

There are three basic tenets of neo-Darwinistic evolution:

- Random genetic variation provides the raw material on which natural selection acts as a directing force.
- Evolution is a slow, gradual, continuous process made up of small micro-evolutionary changes.
- The accumulation of small micro-evolutionary changes over a sufficiently long period of time can result in the large-scale macro-evolutionary changes that are necessary for the development of new species.

In contrast, Goldschmidt believed that abrupt, rapid changes in the germ line generated new species. The germ line consists of the cells that produce the sexual gametes, the sperm and eggs used in reproduction. Sperm and egg cells are haploid; they contain only one copy of the genome (the collection of genes that determine an organism's makeup). When the haploid sperm and egg cells combine during fertilization, they form a new diploid cell, the zygote, which has two copies of its genome, one from each parent. The zygote develops into an adult with its own germ line, which produces its own sexual gametes.

Genes are located on chromosomes. Chromosomes are made up of DNA (deoxyribonucleic acid)-protein complexes. DNA is a molecule that can replicate itself and that carries instructions for creating proteins the organism needs. Humans have twenty-three pairs of chromosomes, including two sex chromosomes, for a total of forty-six chromosomes. They receive one set of twenty-three chromosomes from each parent.

During reproduction, chromosomes are replicated. It is critical that exact or nearly exact copies of the parents' genomes are produced so that they can be passed on to the offspring. However, chromosomal mutations occasionally occur during replication of the genome, perhaps in response to environmental stress, and these mutations may produce significant effects in the fully developed offspring. For example, gametes might have extra or missing chromosomes. Most of the zygotes produced from such gametes are aborted spontaneously, but some are not. A common result of an extra chromosome in humans is Down syndrome. Instead of two copies of chromosome 21, these individuals have three. They possess a characteristic facial appearance and slight to severe mental retardation. Turner syndrome, which affects 1 girl born in about every 3,000, is an example of a missing chromosome. Individuals with Turner syndrome have a broad chest, widely spaced nipples, and a somewhat webbed neck. They fail to mature sexually. Instead of forty-six chromosomes, these individuals have forty-five; they have one X chromosome, but lack the other X or Y sex chromosome.

Sometimes, partial chromosomes are produced during replication of the genome. At other times, chromosomes become inverted, which may repress a gene's action. These types of genetic mutations, although difficult to identify, are more likely to result in viable offspring. Successful "saltations" (the biological term for these drastic changes from one generation to the next) are especially obvious in species bred by farmers.

Goldschmidt called the results of these mutations "hopeful monsters," and he believed that they were more likely to spawn a new species than the gradual micro-evolutionary changes accumulating over a long period of time that were envisioned by Darwin.

Goldschmidt's use of the term "monster" implied the offspring were radically different from the parental type. The monsters were hopeful because they aspired to find an appropriate mate that would allow them to breed true so they could pass on their new traits rather than having them die out. The monsters were also hopeful that their new (novel) changes would allow them to either outcompete the parental population or, alternatively, find a new environment appropriate to their novel qualities. Darwin acknowledged the existence of monstrosities, stating that "under domestication monstrosities sometimes occur which resemble normal structures in widely different animals." But he argued that, "if monstrous forms ... ever do appear in a state of nature and are capable of reproduction ... their preservation would depend on unusually favorable conditions."[1]

In the scientific world of evolutionary biology, it was Goldschmidt who came to be viewed as a monstrosity, out of step with the main current of thought and ridiculed as an unfit mutation. Yet he steadfastly believed in his ideas and hoped that one day his theories would be accepted.

There is no logical reason to discard the hopeful-monster concept in its entirety. Saltatory genetic mutations or sudden changes in the gene line often produce "monsters" that survive and breed true. Giantism and dwarfism are common among humans, and these conditions are not so disadvantageous as to prevent reproduction.

Pygmyism is an even more viable condition in humans, and pygmies regard the rest of us as hopeful monsters. To pygmies, "normal" humans are, understandably, perceived as giant, awkward, belligerent, noisy, ugly beings that crash through the forest, scare the wildlife, and generally mess up the environment. More specifically, it seems that pygmies are better runners than non-pygmy humans. In January 1989, a group of scientists went to Bipindi, South Cameroon, in the tropical rain forest of west central Africa, a region where the Bakola Pygmies live. There, in temperatures between 32°C and 34°C, seventeen male Bakola Pygmies were selected at random to participate in a study. They were asked to breathe through a mouthpiece while they walked and ran on a treadmill powered by a generator. The scientists found that the pygmies expended less energy running than other African populations of the same age and less than taller Caucasian endurance runners. Further, they found that the pygmies ran faster than would be expected based on their bodies' capacity to consume and use oxygen.[2] These findings suggest pygmies are mechanically better or more efficient runners than Caucasians.

Because it is evident that such "monstrosities" do in fact exist, the "unusually favourable conditions" that Darwin said would be necessary for them to emerge and survive must also have existed. Such favorable conditions could conceivably include periods when the climate changed suddenly as a result of volcanic eruptions that darkened the skies and temporarily cooled the planet; altered ocean currents and wind patterns that resulted in rapid warming or cooling causing droughts; failure of the monsoons; expanding ice sheets; or rapidly increasing amounts of greenhouse gases in the atmosphere that generated global warming. In these situations, slow evolution may not have generated change rapidly enough for previously "fit" species to adapt and survive.

Biologists have found potential hopeful monsters in lab mice that have an extra copy of genes that keep cell division under control. These "supermice" are extra resistant to carcinogens, which are known to trigger cancer. This is because their extra genes limit the ability of cells to grow out of control and destroy normal tissue. The supermice are otherwise normal in that their longevity and ability to reproduce are unaffected.[3]

In the next chapter we will examine in detail a recent human example of non-Darwinistic evolution in which genetic change occurred over a matter of months in response to a sudden change in the environment.

15

Dutch Hunger Winter Babies

> Development and epigenetic evolution are expressions of the whole organism and its female parent operating in the larger environment, as well as expressions of genes.
>
> *Robert G. B. Reid,* Biological Emergences: Evolution by Natural Experiment

Even though neo-Darwinist adherents have been reluctant to consider the possibility that organisms can undergo rapid evolutionary change in response to a changing environment, researchers brave enough to look have discovered some very interesting results. The Dutch Hunger Winter of November 1944 is one example of a short but significant environmental event that altered the regular developmental process and led to rapid developmental changes that could potentially have direct evolutionary consequences.

In the last months of 1944, as Allied forces fought to liberate the German-occupied Netherlands, the Nazis imposed a blockade that prevented the transportation of food over inland waterways to the western part of the country, including the cities of Amsterdam, Rotterdam, and The Hague. By early November 1944, an exceptionally harsh winter had set in, freezing the canals and making them impassable for barges. Although the embargo had been partially lifted, food stocks in western Netherlands were quickly depleted. Official food rations fell to below 1,000 calories (4,200 joules) per person per day by the end of November (note that the recommended daily amount for adults is around 2,000 calories or 8,400 joules) and remained so until the end of April 1945. The situation was so grave that the population is said to have eaten tulip bulbs provided by the government. The famine ended in May 1945 when, with liberation, Allied food supplies arrived in the region.

During the Hunger Winter, as many as 30,000 people died of starvation; many of those who survived developed diabetes, heart disease, and cancers. Scientists subsequently discovered that women who had experienced the famine as first- and second-trimester fetuses within their mothers' wombs subsequently bore babies with abnormally low birth weights. This continued for several generations: the descendants of women born during or immediately after the Hunger Winter also had babies with abnormally low birth weights, even though pregnant and nursing mothers in ensuing generations were receiving enough calories in their diets.

According to L. H. Lumey, an epidemiologist, these low birth weights continued to appear because the DNA methylation patterns of the starved mothers had been altered, and their babies had inherited these altered patterns.[1]

As described in "Giants and Pygmies," each cell in our body contains a complete complement of DNA. Unlike simple organisms like bacteria, which repeatedly reproduce carbon copies of their cells, complex organisms like humans can develop differentiated cells that perform specialized roles. For example, human nerve cells transmit electrical signals and make neurotransmitter molecules, but they have lost the capacity to digest or detoxify food. Liver cells synthesize detoxifying enzymes and bile but no longer produce digestive enzymes. Some genes are "shut off" and others are "turned on" in order to make such specialization possible. During embryonic development, some genes can be repressed, but they might be turned on again if circumstances require it. Genes are turned on or off by the normal process of methylation of DNA, which allows cells and organs to specialize and perform specific functions.

Lumey found that the methylation patterns of babies born during the Dutch Hunger Winter were altered, resulting in reduced weight gain and reduced birth weight. When the female Dutch Hunger Winter babies subsequently became mothers, they were more likely to have babies with reduced weight gain and reduced birth weight.

Obviously this kind of result is difficult to replicate scientifically. However, a group of scientists led by Vincent Sollars determined that extreme stress caused by high temperatures, exposure to oxidants, infection, or abnormal metabolic rates could alter normal methylation patterns.[2] Researchers suggest the alteration is caused by a molecule called a heat-shock protein. Heat-shock proteins normally help maintain basic cell functions, like keeping a constant temperature and a chemical balance in the cell. However, when the cell is under stress – for example, when it is exposed to increased temperature, oxidants, or infection – heat-shock proteins undergo a conformational change and activate the production of heat-shock genes. These genes repair damaged proteins and defer the process of programmed cell death or cell suicide (a natural occurrence that can be delayed – to help keep nerve cells alive after injury, for example).

When an organism is under stress, heat-shock proteins can activate genes that are normally shut off or repressed. For example, when fruit flies are stressed, they develop abnormalities, such as bristles that grow in their eyes. Other environmental influences can also affect genome expression. When the diet of laboratory mice was altered, the color of their offspring's coats changed from yellow to brown, and they became less ravenous and prone to cancer and diabetes than their parents.[3] The DNA of the fruit flies and the mice did not change. What did change was the DNA methylation pattern, which permanently shut down some of the genome in an abnormal way. The consequent changes in methylation patterns may persist for many generations, even though later generations are not exposed to continuous or repeated stress.

Such an alteration in genetic inheritance can happen rapidly in response to a changing environment, and it does not require the organism to change its DNA hardware. Rather, it changes only the software that tells the DNA what to turn off. This process is so sensitive to changes in the environment that it can alter an organism's characteristics in a single generation and continue to affect generations far into the future.

Disease is usually the result of infectious agents (such as bacteria, parasites, viruses, etc.), adverse conditions in our environment (for example, famine or secondhand smoke), defective genes we inherited from our forebears (like Huntington's disease or hemophilia), or a combination of these factors. However, as evidenced by the Dutch Hunger Winter babies, the physiology and characteristics of humans – and animals – can be affected not only by their complement of genes and how those genes interact with environmental conditions, but also by the environment's effect on their ancestors.[4] This means that individuals inherit both the DNA, which we can call the hardware, and the methylation patterns, which we can call the malleable software that responds to changes in their ancestors' environment. It also means that individuals can inherit modifications in their ancestors' software without living in the adverse environment themselves. Subsequent offspring are also born with this new set of software instructions.

Even though the DNA hardware does not change, the instructions for how it is used can change quickly and be passed down through the generations.

These observations imply that evolution does not occur strictly through the slow and gradual process of natural selection. The software that dictates how genes are expressed is inherited, and although most known examples of persistent methylation pattern change are detrimental, some changes, like those in the laboratory mice noted earlier, can be beneficial and might provide the opportunity for radical developmental shifts that could produce novel organisms able to adjust to rapid changes in environment. This means that the environment is far more important as a mechanism for rapid evolutionary change than we previously believed. It means that options for rapid change exist. However, because we have been so focused on natural selection and survival of the fittest as determinants of evolutionary change, we have been slow to see other options. If you don't look, you won't see.

Let us look, then, at what the environment holds for humanity, both today and tomorrow.

Today and Tomorrow

16

Today and Tomorrow

What does it matter if you do not believe me?
The future will surely come.
Just a little while
And you will see for yourself.

Cassandra, in the Oresteia *by Aeschylus*

The amount of warming the Earth experiences is directly dependent on the amount of greenhouse gases in the atmosphere, particularly carbon dioxide (CO_2) – a long-lived gas that remains in the atmosphere for decades, centuries, and millennia. Because it lasts so long, it has a long-term influence on climate.[1]

Today's Climate

The amount of CO_2 emitted into the atmosphere has increased every year since 1990, with the exception of 2009, when yearly emissions decreased by 1.3 percent. This exception was in large measure a result of a contraction in global production, capital investment, and consumption. Yet even though emissions in 2009 were lower than those in 2008, they were still the second highest in human history, just behind 2008's total of 9.45 billion metric tonnes. As we send more and more CO_2 into the atmosphere, the atmospheric concentration of the gas rises. In 1990, the amount of CO_2 in the atmosphere was 354 parts per million (ppm). By August 2011, it had reached 390 ppm.[2] Although plants, which use CO_2 to live and grow, do not mind this extra CO_2 in the atmosphere, all oxygen-breathing organisms, including humans, have a little less to breathe. However, the critical problem with rising CO_2 levels in the atmosphere has to do with their influence on climate.

Earth's atmosphere consists of 78 percent nitrogen, 21 percent oxygen, and smaller amounts of several greenhouse gases, including water vapor, CO_2, methane, nitrous oxide, and others. Greenhouse gases act like glass in a greenhouse; they allow visible light (the shorter-wavelength radiation from the sun) to reach Earth's surface, but they trap longer-wavelength infrared energy emanating from Earth and prevent its escape into the upper atmosphere. In other words, CO_2 and other greenhouse gases trap heat next to Earth, much like the glass traps warm air inside a greenhouse and helps the plants grow. The more CO_2 in the atmosphere, the more heat is trapped and the warmer the surface of Earth becomes.

This natural process is what keeps Earth warm enough for us to live on. CO_2 and methane increase and decrease naturally as Earth progresses through ice ages and warm interglacial periods. During warm cycles, permafrost melts and releases methane and CO_2 that were previously frozen in the ground. Changing ocean productivity also influences the amount of CO_2 in the atmosphere. When the oceans are more productive, phytoplankton consume more CO_2. This causes more CO_2 to be taken out of the atmosphere.

These and other natural forces have caused the amount of CO_2 in the atmosphere to vary over the last 800,000 years, from about 180 ppm during ice ages to 280 ppm during warm interglacial periods. Our current level of 390 ppm far exceeds anything that has naturally occurred over at least the last 800,000 years (see Figure 1.5). The amount of CO_2 and other greenhouse gases in the atmosphere is currently rising because of human activities – primarily our use of fossil fuels and our agricultural and other activities that change the makeup of the land. Emissions from fossil fuels and cement make up 88 percent of humanity's carbon emissions; land-use changes make up 12 percent. Coal burning now contributes more CO_2 emissions than the burning of other fossil fuels does.

As our CO_2 emissions rise and the coincident atmospheric CO_2 levels increase, the average surface-air temperature also rises around the world. In 2005, the average surface-air temperature on Earth was about 1°C higher than it was in 1800, just prior to the Industrial Revolution. However, in some regions of Asia, Africa, southern Europe, and North America, the average temperature increased by 1°C to 2°C between 1970 and 2004.

The impacts of these higher temperatures and increased atmospheric CO_2 levels are already being felt:

- Glaciers are melting around the world.
- Permafrost is melting and becoming increasingly unstable.

- Global average sea levels have risen by an average of 3.2 millimeters per year between 1993 and 2011.[3]
- There have been heavy rainfall events – for example, in July 2005, there was a rainfall of 94.4 centimeters in some parts of Mumbai, India's financial center, stranding 150,000 people in train stations, closing airports, and breaking communications linkages. Before that, the heaviest single rainfall was 83.82 centimeters on July 12, 1910, at Cherrapunji, India, one of the rainiest places on Earth. Scientists expect these events to increase in frequency and size.[4]
- In the northern hemisphere, the amount of snow cover that persists through March and April has been reduced.
- Many people in large cities are dying during increasingly intense heat waves – for example, 35,000 died during the European heat wave of 2003, which caused US$13.5 billion in direct costs.
- Droughts have intensified – during the 2006–7 summer season, Australia experienced the worst drought in 1,000 years, and the 2011 drought in East Africa contributed to widespread famine for 11 million people as food prices and poverty soared.
- The geographical distribution of species has changed as plants and animals move poleward and/or to higher elevations in search of a cooler climate.
- Some species have become extinct, including the golden toad of Costa Rica; the Baiji dolphin of China, the "goddess of the Yangtze," a species that had existed on Earth for 20 million years, is considered possibly extinct.

Spring is coming earlier, leaves are unfurling sooner, and birds are migrating and laying their eggs earlier. Fish are migrating up the rivers to their spawning grounds earlier than they have done before. In the Arctic, where temperatures are increasing at a rate of 0.46°C each decade, change is already occurring much more rapidly than elsewhere around the world. Tundra and sea ice are diminishing, which causes serious repercussions for Arctic fauna and flora, including polar bears and migrating birds. For humans, traditional hunting and travel over snow and ice has been affected; Inuit people in Canada's north have been forced to move their villages to higher ground. Valuable timber and habitat of forests in Canada and the United States are being lost as the number of forest fires and pest infestations increase, the result of warmer and dryer summers, milder winters, and human interaction with forests. Other apparent changes include the onset of warmer and dryer conditions in Sahelian Africa (the grasslands south of the Sahara Desert), with a detrimental impact on the area's crops, and the loss of wetlands and mangrove swamps.

The Earth's climate is influenced by the level of CO_2 in the atmosphere and by the oceans' response to its changing levels. The oceans are the greatest sink for greenhouse gases, including CO_2. This means they absorb CO_2 out of the atmosphere and cause atmospheric CO_2 levels to drop. Oceans also assist in the transfer of heat from warmer low latitudes to colder high latitudes. In the Atlantic Ocean, this heat transfer occurs through a mechanism called the North Atlantic meridional overturning (AMO), which operates much like a conveyor belt, bringing warm water from the equatorial Atlantic up to the North Atlantic, giving up heat to the atmosphere, and drawing cold, dense water, which sinks in the Greenland and Labrador seas, south to the Indian and Pacific oceans as deep currents. Solar heating, rain, and clouds generate global-scale differences in water density at the sea surface. These differences are reduced when the water is mixed by winds and tides. In the past, when the North Atlantic Ocean was cooler and meltwater from ice sheets and reduced evaporation caused it to be less saline, the AMO slowed considerably. On rare occasions, the AMO shut down completely, cooling the mild European climate by up to 8°C. There were other times when the AMO's strength increased, warming sea-surface temperatures by up to 3°C over a period of about 500 years.[5] These examples help us understand the major role oceans play in regulating the Earth's climate.

Because it takes so much longer for the oceans to absorb CO_2 from the atmosphere than it does for humans to put it up there, atmospheric levels of CO_2 are rising far more quickly than the pace at which the oceans can absorb it. (Typically, about 50 percent of the CO_2 in the atmosphere is removed through natural processes within thirty years; a further 30 percent is absorbed within a few hundred years, and the remaining 20 percent over thousands of years.) Today, the oceans are busy absorbing CO_2 emissions generated 30 to 100 years ago. They have barely made a dent in the emissions we have dumped into the atmosphere over the last thirty years. And because we keep dumping more and more, faster and faster, the oceans are getting farther and farther behind. Not only that, but as we dump more emissions into the atmosphere, the oceans are taking longer to absorb CO_2. Ultimately, over thousands of years, all of our CO_2 emissions will be absorbed by ocean and land processes; it will just take time. However, we are now at the point at which the natural absorption processes can no longer keep up. All the CO_2 the oceans have absorbed is beginning to influence climate in a profound way. Even if we stopped all our emissions today, the climate would continue to get warmer for decades and centuries to come as the CO_2 currently in the atmosphere worked its way through the system. New research indicates that once temperatures warm, they will stay warm for more than 10,000 years.[6]

One of the consequences of current climate change is that the AMO is changing. Further slowdowns are predicted as meltwater from the remaining ice sheets flushes into the Atlantic. In addition, melting ice caps and glaciers, along with diminished snowfall, are reducing the amount of solar radiation reflected out of our atmosphere, further contributing to warmer temperatures. It is clear that much more change is to come as the oceans respond to the escalating CO_2 levels in the atmosphere.

As the average temperature rises higher, exceeding 1°C above year 1800 levels, more things will change. Global ecosystems will be transformed, and one-tenth of those ecosystems are expected to shrink by 2 percent to 47 percent. In Africa, crop yields are expected to fall; the richest area of flora in the world, the Great Karoo grass flats in South Africa, is predicted to shrink significantly. In the Kalahari, the dunefields will begin to activate; as vegetation cover declines, winds will blow the crests off the dunes, transporting dust, sand, and other sediments to new locations. As glaciers melt in Peru, there will be a serious shortage of drinking water; this will also affect agriculture and energy production in that country. Similarly, in the Mediterranean and in steppic regions around the world, droughts will generate water stress and crop failure. Other regions of North America, Europe, and Russia will experience increased carbon fertilization as plants absorb more carbon from the atmosphere, helping them grow and generating larger crop yields. Up to 50 percent of the Queensland rainforest in Australia will be lost, endangering endemic frog and reptile species; species inhabiting the Dryanda forest of Western Australia will become extinct. As oceans take up more and more CO_2 from the atmosphere, ocean chemistry will change and become more acidic, resulting in the death of corals.

Within the next few decades

Based on a variety of scenarios, including those assuming reductions in CO_2 emissions, scientists predict that, starting in 2020, global average temperatures will increase by about 0.6°C per decade. By 2029, global average temperatures will have reached about 2°C above preindustrial levels. Warming is expected to be most significant over land

and in the northern latitudes. Temperatures will rise more slowly over the Southern Ocean and parts of the North Atlantic Ocean.

Changing precipitation and evaporation patterns will accompany rising temperatures because precipitation, evaporation, and temperature are interdependent. For example, warmer winters are associated with greater high-latitude precipitation in the form of snow because the warmer atmosphere holds more moisture. Warmer summers mean shorter intervals of snow-covered landscapes and sea-ice coverage. This, in turn, makes the surface of the Earth duller, reducing the amount of solar radiation reflected out of our atmosphere, which further intensifies the warming global climate. As a result of climate warming, precipitation will likely increase in the high latitudes and decrease in subtropical regions. Further, rain will come in more extreme events; frequent and/or intense storms will replace gentle showers.

Nearly one-sixth of the world's ecosystems are going to be transformed by climate change, with anywhere from 5 percent to 66 percent of them shrinking. Many animal and plant species will not be able to survive.

If humans continue to dump CO_2 into the atmosphere, levels of 640 ppm will be reached and temperatures will warm more than 2°C above today's levels. With these changes, the North Atlantic meridional overturning is expected to be reduced by about one-third. The cooler temperatures this brings to Europe will be offset by global warming.

Scientists predict that by 2020, between 75 million and 200 million people in Africa will be vulnerable to water stress as climate change causes countries to draw down their naturally renewable water resources. (When a country uses more than 20 percent of its water reserves, it is said to be suffering from water stress. As a result, its development is limited. Countries that withdraw 40 percent or more of their water reserves are under high stress.[7]) If their reserves are not naturally replenished through rainfall or snowmelt within about two and a half years, their water resources may be permanently depleted. By 2020, yields from rain-fed agriculture could be reduced by up to 50 percent. Throughout Asia, particularly in large river basins, there will be less fresh water, affecting more than 1 billion people. Ironically, these same people, as well as those in Africa and those living on small islands, will be vulnerable to extreme flooding around river deltas and along coastlines. By 2050, high-latitude regions can expect an increase in water availability of between 10 percent and 40 percent; however, more of that water will come as heavy precipitation events that will result in flooding. In dryer regions and the mid-latitudes, as well as in the dry tropics, the water available for drinking, agriculture, health and sanitation, and energy uses is expected to fall by 10–30 percent.

A temperature increase of 4.6°C will likely cause the Greenland and West Antarctic ice sheets to melt, raising the average sea level by between 7 and 13 meters or more.[8] Rising sea levels of this magnitude will flood many of the world's coastal cities, regions that during the twentieth century have experienced considerable growth in population and capital infrastructure investment. A 7-meter rise in sea level would require the relocation or protection of between 10 percent and 17 percent of the world's population. Although it may take hundreds or thousands of years for us to experience the full impact of that melting, it is our actions over the next few decades that will determine whether these ice sheets melt.[9]

The level of warming that the world experiences as we move farther into the twenty-first century will be a direct consequence of the amount of greenhouse gases that we dump into the atmosphere today, tomorrow, and over the next few decades. Even if we stopped emitting CO_2 today, past emissions commit the world to some degree of warming (estimated to be at least another 0.6°C by the end of this century). If we reduce our CO_2 emissions but still emit more than can be removed by natural systems, the level of CO_2 in the atmosphere will continue to rise, but it will do so more slowly.

When CO_2 concentrations reach 550 ppm, temperatures are expected to reach about 3°C higher than preindustrial levels. Scientists agree that when escalating temperatures reach 2°C to 3°C above the average global temperatures of 1850, or when atmospheric levels of CO_2 surpass 450 ppm, numerous damaging effects will ensue (see Table 16.1). Few of the world's ecosystems will be able to adjust to an increase in temperature of 3°C. But even the most "green" scenarios put forward by scientists predicting future climate change show temperatures rising beyond 3°C in the northern latitudes by the end of the twenty-first century. Worst-case scenarios show global temperatures rising beyond 7.5°C.

Where do we go from here?

These fundamental shifts in the climate system remind me of an ocean liner. It takes a long time to adjust the course of such a big boat, but once it starts to change, it is very difficult to stop. Our ocean

Table 16.1. *Global effects of rising temperatures and atmospheric CO_2 concentration levels*

Temperature increase	**Atmospheric CO_2 concentration levels**	**Effects**
1.0 to 3.0°C	380 to 550 ppm	Agricultural yields will begin to fall in Africa. In Peru, there will be shortages of water for drinking, agriculture, and energy. Steppic regions around the world will experience greater drought, generating water stress and crop failure. Some regions in the northern hemisphere will experience larger crop yields. One-tenth of global ecosystems will lose between 2 percent and 47 percent of their extent. The Great Karoo grass flats in South Africa will shrink. The Queensland rainforest in Australia will potentially lose 50 percent of its extent. The Dryanda forest of Australia will experience species extinctions. The Kalahari dunefields will become activated. Oceans will become more acidified, resulting in dieback of corals. There will be a higher incidence of flooding in low-lying areas around the world.
1.5 to 2.5°C	400 to 500 ppm	One-fifth to one-third of plant and animal species will face extinction. The Greenland ice sheet may begin to melt, causing the beginning of an expected long-term 7-plus-meter rise in sea level. Clean fresh water will be increasingly unavailable because of droughts and flooding. Water availability is expected to fall by between 10 percent and 30 percent in dry regions and mid-latitudes. By 2020, between 75 million and 200 million people in Africa will experience water stress, and rain-fed agriculture could decline by as much as 50 percent. In Asia, more than 1 billion people will face reductions in fresh water and be vulnerable to flooding.
2.0 °C	450 ppm	Between 1.0 billion and 2.8 billion people will have trouble getting drinking water. Ninety-seven percent of the coral reefs in the world will die. Sixteen percent of the world's ecosystems will be transformed, with between 5 percent and 66 percent losing some of their extent. Rising sea levels combined with cyclones will force millions of people in coastal regions to relocate.

beyond 2.0 °C	> 450 ppm	The world's cereal crop yield will fall by 30 million to 180 million tons; up to 220 million people will risk hunger. There will be large-scale displacement of people in Africa as a result of poverty, starvation and thirst; millions more people will be at risk of malaria. The total loss of summer ice in the Arctic will mean the destruction of Inuit hunting culture, with severe impacts on walrus populations and a drop of 30 percent in the polar bear population by 2050, a predicted 60 percent decline in the lemming population, and a decline in the stability of the Arctic tundra (less than half of it will remain stable), endangering High Arctic shorebirds and geese. In the Americas, vector-borne diseases are anticipated to expand poleward; more cases of malaria will be seen in North America as the mosquitoes that carry the disease extend their range. Interregional tensions will likely mount in Russia when crop productivity falls as temperatures continue to rise. In Asia, 1.4 billion to 4.2 billion people will face water shortages, and vector-borne diseases will start appearing farther north and south. China is expected to lose half of its boreal forest. Half of the Sundabans wetland in Bangladesh will also disappear. In Australia, the threat of extinction will expand; half of the Kakadu wetlands will be lost.
2.0 to 3.0 °C	450 to 550 ppm	The overturning of the North Atlantic Ocean will be reduced by about one-third.[a] The Amazon rainforest will collapse. Hundreds of millions of people will suffer from increasing water shortages. Between 20 percent and 30 percent of plant and animal species will be at risk of extinction. In Africa, 80 percent of the Karoo grass flats will potentially be lost, endangering 2,800 types of plants; five of South Africa's natural parks will lose more than 40 percent of their animals; fisheries will be lost in Malawi; three-quarters of South Africa's crops will end in failure; and all of the dunefields in the Kalahari will be active, threatening the ecosystems and agriculture of the sub-Sahara region. Maple trees, the source of Canada's world-renowned maple syrup, will be threatened. In Australia, the Kakadu wetlands and alpine zone will be lost. The Tibetan plateau in Asia will undergo desertification and a shift in the permafrost. The Chinese boreal forest could disappear.

(continued)

Table 16.1. *(continued)*

Temperature increase	Atmospheric CO_2 concentration levels	Effects
above 3.0 °C	> 550 ppm	The gross domestic product of sixty-five countries will fall by 16 percent. As malaria continues to expand, so too will dengue fever, with potentially more than half of the world's population exposed to the disease, compared to one-third in 1990.
4.0 °C	700 ppm	Entire regions will be unable to produce agricultural crops, resulting in more people at risk of hunger. Australia will no longer be able to produce agricultural crops and will no longer have an alpine zone. In Africa, 70–80 percent of the human population will face hunger. In Russia, a 5–12 percent drop in agricultural production is predicted to occur over 14–41 percent of the country's agricultural regions. In Europe, 38 percent of the alpine species are projected to lose 90 percent of their alpine range. Malarial regions are predicted to expand by 25 percent. Twenty percent of the perennial zones will disappear, along with more than 40 percent of the taiga-producing regions. The extent of tundra regions will potentially be reduced by 60 percent. Timber production will increase by 17 percent. North Atlantic thermohaline circulation will slow significantly.
2.0 to 4.5 °C	450 to 775 ppm	A 20- to 400-million-ton reduction in global cereal production is anticipated to result in up to 400 million additional people at risk of hunger; 70–80 percent of those people will be in Africa. An additional 1.2 billion to 3.0 billion people will suffer from water stress.

There is a 50 percent chance that the west Antarctic ice sheet will begin to melt; about one-third of global coastal wetlands will be lost.

In the Americas, half of the world's migratory bird habitat will potentially be lost.

Alpine species in Europe and Russia are projected to be near extinction, along with 60 percent of Mediterranean species.

In China, rice yields may fall by 10–20 percent.

Australia is predicted to lose half of its eucalyptus.

Notes: There are many uncertainties inherent in climate-change projections, including the level of future emissions, the climate response to those emissions, future population levels, economic development, technological development, and changes in behavior. These need to be added to climate modeling uncertainties. Scientists have assigned a series of confidence levels to their predictions, from very high (90 percent chance of being correct) to very low (less than 10 percent chance of being correct). Researchers are often wary of making dire predictions because they do not like to be responsible for initiating actions that turn out to be unnecessary; however, most individuals and policy makers should be more worried about hedging their bets and taking precautionary steps in case serious climate change does occur. The events listed in this table may happen at higher or lower temperatures or CO_2 concentration levels, and some are predicted with greater or lesser confidence than others. All are based on published research done by scientists from around the world (see the Sources section that follows). Although there is some question as to exactly what will happen when, there is no doubt that if we fail to alter our attitude and reduce our CO_2 emissions, these things will begin to happen. For a popular reference into descriptions of what each degree of warming would mean for humanity, see Lynas, M. (2008). *Six Degrees: Life on a Hotter Planet*. Margate, FL: National Geographic.

This table was originally published in Hetherington, R. and Reid, R. G. B. (2010). *The Climate Connection: Climate Change and Modern Human Evolution*. Cambridge: Cambridge University Press, pp. 276–79.

[a] Weaver, A. J., Eby, M., Kienast, M., and Saenko, O. A. (2007). Response of the Atlantic meridional overturning circulation to increasing atmospheric CO_2: Sensitivity to mean climatic state. *Geophysical Research Letters*, 34, no. 5, L05708.

Sources: Much of the scientific data presented in this table has been obtained from three main sources:

Intergovernmental Panel on Climate Change (IPCC) (2007). *Climate Change 2007: The Physical Science Basis*. Contribution of Working Group I to the Fourth Assessment Report of the Intergovernmental Panel on Climate Change, ed. S. Solomon, D. Qin, M. Manning, Z. Chen, M. Marquis, K. B. Averyt, M. Tignor, and H. L. Miller. Cambridge: Cambridge University Press, available online at: http://www.ipcc.ch/publications_and_data/ar4/wg1/en/contents.html

IPCC (2007). *Climate Change 2007: Impacts, Adaptation and Vulnerability*. Contribution of Working Group II to the Fourth Assessment Report of the Intergovernmental Panel on Climate Change, eds. M. L. Parry, O. F. Canziani, J. P. Palutikof, P. J. van der Linden, and C. E. Hanson. Cambridge: Cambridge University Press, available online at: http://www.ipcc.ch/publications_and_data/ar4/wg2/en/contents.html

Warren, R. (2006). Impacts of global climate change at different annual mean global temperature increases. In *Avoiding Dangerous Climate Change*, eds. H. J. Schellnhuber, W. Cramer, N. Nakicenovic, T. Wigley, and G. Yohe. Cambridge: Cambridge University Press, pp. 93–131.

liner Earth has been changing direction for some time now. We just have not been watching from the bridge. Now that we have climbed up to take a look, we are discovering that things are happening far more quickly than we thought possible. Our ship has changed direction and its course is difficult to alter. Even though these changes are happening over decades and centuries, they are occurring too quickly for organisms, including humans, to evolve to deal with them. Natural selection and slow gradual evolution will not help us deal with the sudden changes appearing on our horizon.

What level global temperatures reach, and when, will depend on whether our global CO_2 emissions continue to escalate, are stabilized, or are reduced. To predict future climate changes, scientists have generated a set of scenarios based on different projections of future economic growth and population, future human behavior, and the development and adoption of more efficient technologies,[10] all of which will determine the amount of CO_2 emitted. In most of these scenarios, atmospheric CO_2 levels are expected to exceed 600 ppm after 2050.[11]

If we reduce our CO_2 emissions, the world will reach atmospheric CO_2 concentrations of 550 ppm somewhat later than it will if we do not reduce emissions. Such a delay may give us time to change our technology, behavior, and culture and give ecosystems a chance to adjust to the changing circumstances.

To maintain atmospheric CO_2 concentrations at 450 ppm, which would imply a warming of about 2°C, we would have to reduce our emissions by about 90 percent by 2050.

There is no question that change is imminent. The question remaining is whether we embrace change or have change forced on us – by Nature.

Embracing change requires that we accept it is coming and acknowledge our responsibility for that change. This is an exceedingly difficult action, particularly for those people and organizations that currently appear to be successful. Success, particularly when it has been achieved through the fossil fuel economy, gives people a sense of, if not actual, dominance – a feeling that we are doing things the "right" way. As a result, it is hard for us to admit that it is in our best interest to change our way of doing business. The theories of competition, survival of the fittest, and natural selection allow us to believe that because we are successful now, we will always be successful. Because we are dominant, whatever we have done to get us here must be right, will always be right, is the only way. But remember Albert

Einstein's wisdom: "We can't solve problems by using the same kind of thinking we used when we created them." It is time to let go of old ways of doing and look to new ways of being.

In the past, when things did not go well on the land, we turned to the massive expanse of the world's oceans to feed our people, dump our sewage, test our technologies, or simply take us somewhere better. Yet, as the next chapter indicates, it appears even these vast, untamed expanses have been severely affected by our wanton ways. There is no New World to discover. We have reached Earth's tipping point and are clinging to the hope that some gradual natural phenomenon will save us. However, it is not something gradual that is required, but rapid action – our own.

17

Dead Zones

> To understand and manage the oceans is to understand and manage ourselves.
>
> *Peter Keller, 2007*[1]

I spent my childhood summers combing the beaches and playing in the fields and mountain meadows of Salt Spring Island on Canada's west coast. Both sets of my grandparents lived there in country cottages with expansive gardens. My great-aunt and -uncle and some other relatives by marriage also resided on the island. My father's sister and her husband and my four cousins lived on a sheep farm just down the big hill from the house my grandfather built at the base of Mount Maxwell. I envied them.

Salt Spring was relatively undiscovered then. My brothers and sisters and I would catch the ferry from Tsawwassen, near Vancouver, and feel the city's clutches slipping away as we crossed the Strait of Georgia, often reaching "our" island under cover of night. The darkness didn't matter as I could find my way along the single road blindfolded.

It was here one summer in my early twenties that I attended the marriage of my cousin. Weddings on Salt Spring were wondrous things; we knew almost all the kids who lived on the island, and these social "barn dance" events were often accompanied by bonfires on the beach that lasted until the wee hours of the morning. I best remember my cousin from years earlier when her family and ours had a great summer bonfire on one of the island's many beaches, which were then usually deserted. We ate oysters that we had plucked from the shallow shores that afternoon and shucked while we sat warming our toes and fingers by the fire and gazing up at a star-filled sky. As an awe-struck nine-year-old, I remember my cousin throwing back her long flowing dark hair and swallowing raw oysters one after another, then laughingly challenging us to do the same.

Her husband is a fisherman, a good and successful one. I have listened in admiration to his tales of fishing halibut, the huge flat fish that can destroy a boat if it is not shot immediately after being caught. About twenty years ago, he began fishing what was considered an infinite supply of hake off Canada's west coast, catching it for an expanding international market. Fishing was so successful that he and my cousin were able to purchase a large fishing vessel, hire employees, and grow a profitable company. They invested their ample returns in real estate and other ventures.

My father loved to fish; perhaps that is where I developed such a knack for it. My first husband and I built a sailboat, and one summer we set sail for Rivers Inlet, a few days' journey up the west coast from Vancouver. We stayed there for weeks. The boat had a full galley, heater, large water tank, and lots of fuel and supplies. Each evening after a full day of fishing, I would crank up the oil stove and can jars of fresh-caught sockeye and coho salmon in the pressure cooker. We would eat a big pot of fish chowder that I made with anything left over.

I became so adept at catching fish that the guides for the fishing lodges used to follow us about, inquiring, "How many pulls today?" I always told them how many pulls of fishing line I had released behind the boat; there seemed an endless supply of fish to go around. We would laugh with delight as we landed one great salmon after another, some weighing more than 23 kilograms.

As the years progressed, I began to notice that I was getting "skunked" at my favorite fishing holes around the Strait of Georgia. Fish were disappearing from places where I had previously always caught my limit. When I mentioned the decline in salmon stocks to my cousin-in-law and asked him whether the same thing could happen to the hake, he guffawed and said with great confidence, "There is so much hake out there that it will never be fished out."

Who was I to disagree? I had seen with my own eyes the plethora of fish beeping across the screen of my cousin's depth sounder when he fished. It really did appear that the abundance of the ocean was limitless.

I also recall my high school geography teacher telling the class, "We will never run out of food. The oceans are so big and have so many fish that all we have to do is harvest the oceans and we will have an endless supply of food for humanity." Relieved that there was such a simple solution to the world's hunger problems, we discarded our niggling concerns, sat back, and relaxed.

And why shouldn't we? One look at a globe makes it clear that oceans cover about 70 percent of the surface of the Earth, more than twice the area covered by the land on which humans reside.

But that was back in the early 1970s, when the ocean's resources seemed endlessly capable of feeding a population of about 3.5 billion people. Change crept up on us – we barely noticed its nuances. Within slightly more than 30 years, 3.5 billion people became 6 billion people, and pollution and red tide stopped my harvesting oysters from Salt Spring beaches. I began to buy locally farmed oysters from the fish market. My cousin stopped fishing hake. The coastal stock was overfished in 2002, although it was subsequently "rebuilt," and the Puget Sound and Strait of Georgia stocks are now considered "species of concern."

Collapse of the cod fishery

And then, in 1992, the once extraordinarily abundant cod stocks off the coast of Newfoundland collapsed. On July 2, 1992, John Crosbie, a Newfoundlander and the Canadian federal minister of Fisheries and Oceans, announced a moratorium on fishing northern cod off Canada's east coast. Two years later, a wider moratorium was imposed on cod fishing across the Grand Banks. We had overestimated both the size of cod stocks and the level that could be sustainably harvested. Factory freezer trawlers, advanced nets, and electronic fish-tracking devices that could find fish with ease dramatically increased the amount of fish that Canadian and international fleets harvested. Overexploitation was inevitable. The collapse and moratorium destroyed the economic foundation of Newfoundland communities.

In February 2007, I received a media alert: "Warming climate, cod collapse, have combined to cause rapid North Atlantic Ecosystem changes."[2] A Cornell University oceanographer reported that the salt water of the northwest Atlantic shelf (an area that stretches from Greenland to North Carolina) is becoming less salty as more precipitation and more fresh water from melting Arctic ice flow into the North Atlantic. At the same time, Arctic wind patterns are shifting, possibly as a result of climate change, and these changing wind patterns are affecting the Atlantic currents. These alterations have made the sea water more like fresh water, and this is having an effect on animal life in the ecosystem. The population of plankton is exploding, partly because its former predator, cod, has disappeared, but also because of this freshening of the sea water.

In September 2010, we heard news that the cod population southeast of Newfoundland had grown by 69 percent over 2007 levels. It was a heartening sign of recovery; unfortunately, however, it still meant that cod levels were only 10 percent of their 1960s levels, and the fish that remain are much smaller than their predecessors.

The appearance of "dead zones"

It is not just the North Atlantic cod that seem to be struggling. Scientists have noted that "dead zones" in the ocean are increasing in size and number.[3] Located predominantly along the continental shelves of populated regions, these zones have low levels of oxygen and are virtually devoid of living things. During the last fifty years, there has been an exponential increase in the number of these oxygen-starved zones.

- Where the Mississippi River flows into the Gulf of Mexico, a large dead zone develops between May and September each year, covering as much as 22,000 square kilometers. Nitrogen fertilizers used on the agricultural land of the Mississippi watershed drain into the river and are believed to contribute to this phenomenon.
- Off the coast of Oregon, a 3,000- to 5000-square-kilometer dead zone now appears for a few months each summer. It appears to be getting thicker, expanding northward, and lasting longer. There is not enough agriculture along Oregon's coast to explain this dead zone, but scientists think it may be linked to falling levels of oxygen in waters farther off Oregon's coast. These huge oxygen-minimum zones are located beyond the continental shelves of the world and are expanding and losing oxygen. They now cover about 30 square kilometers – about 8 percent of the world's oceans.
- On the West Indian shelf, a dramatic change and reduction in fisheries, particularly the shrimp fishery, has been blamed on a recently discovered shallow-water dead zone up to 20 meters deep. This dead zone appears when large amounts of nitrogen are drawn down from the atmosphere into the Indian Ocean.

It is not clear whether these dead zones are becoming larger and more common because of climate change, shifts in ocean currents, pollution, some unknown reason, or a combination of factors.

Climate and related impacts

It *is* clear that the oceans are becoming warmer, and this warming affects marine life. Oxygen dissolves less easily in warm water than in cold, so the organisms that need oxygen have more trouble getting it in warm water than they do in cold water. Animals with larger bodies are especially sensitive to reductions in oxygen. Thus, scientists indicate that if ocean temperatures warm to 22°C, 25 percent of marine organisms will die. If temperatures reach 26°C, nearly all will die. Observations show that the global ocean is warming to depths of up to 3,000 meters. Between 1950 and 2000, the average global ocean temperature rose 0.5°C.[4] However, in some locations, such as the ocean around Japan, sea-surface temperatures increased by more than twice this number.[5]

According to Dr. Daniel Pauly, recent director of the Fisheries Centre at the University of British Columbia, as harvesting has intensified and as the oceans have become warmer, the total number of fish caught in the world each year has been dropping, and the size of the fish caught is steadily declining. There has been a loss of fish diversity and habitat, and the number of non-harvestable creatures, like jellyfish, is increasing. This has been most notably observed by beachgoers on the Mediterranean coast.[6]

Ken Caldeira of the Carnegie Institution at Stanford University has said that "the rate of increase in CO_2 is 100 times more rapid today than during natural glacial/interglacial cycles."[7] The amount of CO_2 that humans are currently emitting into the atmosphere is 70 times the amount that is emitted through natural processes, and scientists believe that humans may eventually emit CO_2 at a rate 200 times greater than natural emission levels. Whether it is 70 times or 200 times, this rate of emission far exceeds Earth's natural capacity to absorb CO_2.

Ocean acidification

One of the major consequences of rising atmospheric CO_2 levels is the anticipated rise in acidity of the world's oceans as they absorb more and more CO_2. High ocean acidity levels limit the ability of marine animals, particularly shellfish, foraminifera,[8] and corals, to produce their hard calcium-rich shells and skeletons; it basically corrodes them. It also impacts their physiology; when CO_2 enters their bodies, it causes their pH to decrease. And it reduces their ability to transport oxygen.[9] Ocean acidification is not new. During the Earth's previous five mass-extinction events, all reef-building ceased; each was associated with changing levels of atmospheric CO_2. For example, around the time of the Cretaceous-Tertiary extinction event 65 million years ago, there was a sudden acidification of the ocean that caused the extinction of almost all carbonate-making species (i.e., corals and other creatures with shells). It took about 4 million years for the oceans to recover.

Today, 20 percent of coral reefs have already died and 35 percent more are seriously threatened.[10] By 2060, when CO_2 levels are expected to have risen to twice the preindustrial levels, very limited coral habitat will remain. Because coral reefs grow in nutrient-poor tropical waters, they provide critical habitat for a diverse array of tropical fish and invertebrates. Without the reefs, these organisms will also disappear. By 2099, if CO_2 levels in the atmosphere continue to rise, ocean acidification will reach levels that will eliminate nearly all habitats in which corals could survive. This is true even under the most optimistic "green" scenario predicted by climate scientists.

Options and actions

The situation appears hopeless. What can we do? It is tempting to bury our heads firmly in the sand and remain ignorant of our actions, but that is indeed hopeless. Rather, we need to recognize that we are disturbing our planet and its species, and we need to take responsibility and positive action. Climate change caused by human activities is fundamentally changing the chemistry of the sea, and overfishing and environmental degradation are exacerbating the profound ecological impacts. The fishing industry receives annual subsidies of about US$30 billion; this encourages overfishing, which wipes out species – and the species that depend on them.[11] Bottom trawling destroys productive seafloor habitat around the world and leaves behind suffocating mud. It depletes the large bottom-dwelling fish, which are also most affected by expanding dead zones. Trawling at any depth collects bycatch – fish and animals that have no commercial value, which are caught and then thrown overboard, dead or dying. In the 1990s, an astonishing one-third to one-quarter of all the world fisheries catch was bycatch.

To counter these attacks on Earth and its oceans, we need to be aware of the demands we are placing on our oceans and their fisheries. We need to ask questions about the food we eat: How is it caught? Is it overfished? Is it caught or farmed in such a way that it harms other marine life or habitat?[12] If it is farmed, is it fed other fish that could be used to feed humans less expensively?

Aquaculture has been heralded as the solution to feeding the expanding global population,[13] and between 1973 and 2005, there was a ninefold increase in aquaculture production, mostly involving carp in China and salmonids elsewhere. However, carp and salmon are fed fish meal made with herring, a cheap food source that can be used to feed many more people than the salmon or carp feed. Ironically, the carp and salmon fed on herring are far too expensive to feed the poor. A few cheap programs for oyster and giant clam cultures in developing countries have proved that aquaculture can feed many people,[14] but we must not forget that aquaculture is a business; its objective is to make money. It may create jobs and reduce the poverty of some individuals, but feeding the world's hungry is not its purpose. Further, aquaculture frequently places increasing demands on the environment and on water resources that will only become more valuable as global warming escalates.

We cannot afford to waste the world's fisheries. We should not buy or kill anything that we do not need or will not use, and we need to make better use of what we have. Dr. Daniel Pauly of the UBC Fisheries Centre suggests that we need to return to eating smaller fish – for example, anchovies.[15] By doing so, we will be using the oceans' resources more efficiently. We need to develop and enforce policies that eradicate the fishing practices that are destroying the oceans' capacity to support life.

In our own neighborhoods, we need to be aware that sewage, agriculture, and aquaculture have important downstream impacts. We have an obligation to our children to recognize the consequences of the actions that we perform or condone. We need to recognize that it is not the fisheries that need managing, but ourselves. We must stop eating from our grandchildren's plates.

One way to facilitate these objectives is to fund and support interdisciplinary research and education that links economic and human behavior with the ecological and environmental impacts of that behavior. Another is to improve our understanding of economics and how it influences our behavior. We must also pay more attention to the role governments play in our lives. The next section delves into economics, governance, and some of the similarities between the underpinnings of economic theory and evolutionary theory. Like Earth's biological cycles that ended in mass extinctions, businesses, economies, and societies also experience cycles that can result in their demise.

The Economic Connection

18

The Economic Connection

If something is good for the corporation, it's per se held to be good for the country.

John Kenneth Galbraith, quoted in a 2003 interview with Stephen Bernhut that appeared in the Ivey Business Journal

A culture defined by imbalance and fragmentation perpetuates itself - keeps us buying - by directing our attention outward. The single greatest threat to the corporate-consumer complex is that on our way to the mall, we might get distracted.

Carol Lee Flinders, Rebalancing the World

Capitalism and democracy - an introduction

Democracy in its original sense of "rule by the people or government in accordance with the will of the bulk of the people" was not a popular notion until the advent of the First World War. Prior to this, democracy was rejected by the ruling elite as a bad thing that gave too much power to the common people, making it "fatal to individual freedom and to all the graces of civilized living."[1]

Canadian economist C. B. Macpherson, in his book *The Real World of Democracy*, illustrates how our present-day liberal democracies began as liberal societies that later adopted democracy. The predemocratic liberal society of the Western world operated with competition and freedom of the market. Both government and society as a whole were organized on the basis of freedom of choice, freedom of religion, freedom of employment. People were free to offer their services and labor on the open market, free to spend their earnings and to invest based on current prices - and in doing so they determined what would be produced and at what price. In this way the resources of the society, its "capital," were allocated as efficiently and effectively as possible. This was the basis of Adam Smith's "invisible hand," which kept the market free from interference and allowed for the most efficient distribution of resources throughout the economy.

In his book, *The Wealth of Nations*, Smith outlined how the combination of technological efficiency and labor could create surpluses that are accrued as profits. "As every individual, therefore, endeavours as much as he can both to employ his capital in the support of domestic industry, and so to direct that industry that its produce may be of the greatest value; every individual necessarily labours to render the annual revenue of the society as great as he can."[2] This system of employing other people's labor and building capital ultimately developed into the capitalist market system, in which capitalists accumulated capital in the form of money, property, businesses, or other wealth. Unlike previous societies, which allocated resources based on custom, status, and authority, the capitalist market society was based on contract and the freedom of the individual to make choices. Where people had previously considered themselves part of communities, ranks, or orders, they now considered themselves to be individuals. In doing so they lost some of their customary security, but they gained the freedom of individual choice. People now had the opportunity to develop new skills that could raise their standard of living.

This new capitalist market society needed a government to help it operate. Revolutions in seventeenth-century England, eighteenth-century America, and eighteenth- and nineteenth-century France led to the establishment of governments that were responsible for supplying political goods and services: laws, regulations, and taxes were put in place that allowed the market society to work; services were provided to defend it; and a military was established to expand it. Services were also provided to educate future workers, provide sanitation, and help industry (through the provision of tariffs and grants). Control of the government was put into the hands of men who were democratically elected among persons of substance. Individuals were free to create, and associate with, different political parties. This was the beginning of the liberal state and competitive political parties.

The role of this new government was "to maintain and promote the liberal state," while "the job of the competitive party system was to uphold the competitive market society."[3] However, pressure built over time as those men with little or no substance began to realize that they had no political influence because they could not vote. And so, with great agitation and organization, came the democratic franchise. For women it came much later, after they had "moved out from the shelter of the home to take an independent place in the labour market."[4]

According to Macpherson, liberal democratic states were developed "to provide the conditions for a competitive, capitalist, market society ... [which] necessarily contains a continuous transfer of part of the powers of some men to others. It does so because it requires concentrated ownership and control, in relatively few hands, of the capital and resources which are the only means of labour for the rest."[5] As Adam Smith wrote, the combination of labor, capital, and technology work together to develop the surplus benefits to the investor. To participate in the market, wage earners transfer some of their power and wealth to the capitalist, which allows the capitalist enterprise to continue to produce. However, this also leads to a widening gap between the rich and the poor, and it can reduce a worker's sense of self-worth if he or she is unable to find a job, secures a low-paying job, or believes social success depends on acquiring wealth.

Despite its drawbacks, proponents promote the theory that the market system ensures the best use of society's resources through competition. Those who are best, fittest – for example, most efficient – outcompete those who are less fit, and, as a result, scarce resources are allocated in the most efficient manner possible. Further, the market society's continued survival and growth requires and depends on its participants' developing an unlimited desire to acquire and consume. This helps keep the market growing and ensures that, despite intermittent slowdowns, the market always self-corrects with further growth. The ups and downs of business cycles are recognized as part of a typical self-correcting business rhythm. Nothing needs to be done to alter this cycle because market forces will ensure its recovery.

When Adam Smith wrote *The Wealth of Nations*, he was considering the domestic economy of Britain, without the influence of globalization. Britain has since undergone substantial economic changes that have seen the mass exodus of its manufacturing sectors and the development of a strong financial sector in London. The present liberal democratic capitalist society is a consumer-driven free-market system in which mass-produced products and services are promoted to an ever-expanding local and international market. Darwinistic concepts of competition, natural selection, and survival of the fittest still form the theoretical basis of local and international market systems. However, while Adam Smith understood the domestic economic system, he did not anticipate the advent of globalization, international financial markets, and overwhelming international trade. It was John

Maynard Keynes, 150 years later, who argued that while free markets are imperative, there must also exist some limited government involvement to provide stability and control of the market.

Increasingly, dominant multinationals compete for and control what are, generally, relatively stable markets, generating wealth for their shareholders and executives. Whereas society and the market were meant to benefit from competition resulting from the most efficient use of scarce resources, now the emphasis is increasingly focused on maximizing the benefit for members of corporate organizations, including executives, managers, and traders. Their personal wealth, gained through high commissions, bonuses, and salaries, has become the focus, putting investors, shareholders, and society at risk.

Recession and depression – and the major threat to global market stability

While business cycles are natural, one phenomenon of the more developed global marketplace is that the effects of both an up cycle and a down cycle are felt more broadly. Economists have become keenly aware that the major threat to the capitalist economic system is recession and depression – periods in which business activity experiences a serious "readjustment."[6] Prices fall; people lose their homes, businesses, and jobs; investors lose their savings; and bankruptcies escalate. To avoid such hardships, we are encouraged to "grow" the economy by consuming more. This does not cause us too much consternation as it is consistent with our competitive survival-of-the-fittest logic. We believe that being the best or fittest means having the most money and the most stuff. As a result, consumers compete to consume the most – the biggest houses, the newest cars, and the latest gadgets – even though this frequently means we are consuming far beyond our means and needs. And we now know that this strategy too can backfire.

In 2003, in his book *Al Qaeda and What It Means to Be Modern*, John Gray presciently stated, "The global free market lacks checks and balances. Insulated from any kind of political accountability, it is much too brittle to last for long."[7] Not only does the new global market feel the effects of an up or down cycle more broadly, but, because of its global nature, it lacks any governing body charged with its oversight that can allay those effects or attempt to avoid them. Gray lays most of the blame for these weaknesses on American policy makers: "By doing all they could to project the free market throughout the world,

American policy-makers ensured that its inherent instabilities became global in scope."[8]

It did not take long for Gray's prediction to come to fruition. On Monday, September 15, 2008, the world awoke to find that Lehman Brothers Investment Bank had filed for Chapter 11 bankruptcy in New York. It remains the largest bankruptcy in world history. Within twenty-four hours, global financial markets came to a complete standstill; all lending and borrowing ceased. The world's largest corporations no longer had access to the daily borrowing they required for their day-to-day operations such as payroll and inventory. What ensued was a global financial crisis. Greed had usurped the market's ability to ensure the best use of society's resources through competition.

The global financial crisis - what went wrong

A basic tenet of the market economy is consumer confidence. Consumers must believe in the viability and strength of their market and financial institutions for those institutions to persist. People must accept and adhere to market and financial system rules for the system to work. The same principle applies to the rule of law and governments. Without acceptance and confidence, the system collapses. For instance, if, on any given day, a large enough group of customers lose confidence in the solvency of financial institutions and withdraw all the cash from their accounts, a "run on the banks" will ensue. If the group mentality pervades deeply and broadly enough, people's behavior will ensure their greatest fear materializes. This is because financial institutions never have on hand all the cash that has been deposited in their accounts. Further, the global nature of financial institutions and markets, combined with recent advances in technology, means that people can communicate their ideas or concerns more rapidly than ever before. Therefore, a local or regional loss in consumer confidence has a much greater chance of becoming a highly contagious global phenomenon. Today, more than at any previous time, we face the possibility of a run on the banks, or any other mass psychological movement, going viral.

According to Alan Greenspan, chairman of the U.S. Federal Reserve Board from 1987 to 2006, another essential ingredient of the competitive free-market system is that "each individual economic entity works assiduously to preserve its solvency."[9] If they do not, according to Greenspan, the system "will not work."

Another basic principle of the market economy is that investors demand higher returns for investment options that have more risk and are willing to accept lower returns for investments with less risk. They are also typically willing to pay more for lower-risk investments because those investments are more likely to yield a return, whereas higher-risk investments may result in a default. Nevertheless, the best option, and one that businesses constantly seek, is a deal that pays a high return with little or no risk.

Traditionally, business has been about relationships. For example, a borrower, such as a couple seeking a home mortgage, goes to a bank and negotiates with a mortgage broker to secure a loan. The lender assumes the risk that the borrower might fail to repay the loan – in the case of the couple buying a house, the lender assumes the risk they might default on their mortgage. As a result of the lending process, a relationship develops between the lender and the borrower.

Recently, however, the traditional lending system evolved into a four-party system involving the borrower (i.e., the homeowner), the lender (the wholesale mortgage lender), and an investment bank that buys mortgages from mortgage lenders and then packages and sells investments to the fourth party, the investor – an individual or institution that purchases securitized investment products.[10] The role of securitization is to minimize the risk to the investor by combining high-, medium-, and low-risk investments and distributing the risk throughout the market. These combined investment products are rated according to how risky they were before they were sold. Investors are willing to pay more for investment products that are less risky, so it is in the best interest of the investment banks to have high-rated, less risky securitized investment products. Higher ratings equate with higher yields.

In this process, relationships become less important – investors know nothing about the borrowers or lenders involved with the investments they are buying, and this is the Achilles heel of the global banking system, which validates John Gray's concern about the lack of checks and balances in today's free-market system. Investment banks and investment bankers, who operate on commission, make more money on investment products with higher ratings. Investment agencies such as Fannie Mae and Freddie Mac are responsible for rating securitized investment products. The best rating that can be assigned is AAA, something given only to the best-performing and most secure companies and organizations in the world, such as the government of Canada or Royal Dutch Shell. From AAA, ratings progress downward to investment gradings AA, A, BBB. The next set of gradings, BB, B, CCC, CC, C, and D ratings, are considered non-investment-grade ratings because they have a much higher risk of default. There is a substantial difference between non-investment-grade and investment-grade ratings, because only investment-grade products may be purchased by regulated institutional investors such as insurance companies, pension funds, and commercial banks. Ratings agencies make more money rating investment products as AAA than CCC, because it is easier for the investment banks to recommend these and place them in the portfolios of their investors. Higher ratings result in higher turnover and greater demand.

Initially, investment products were packaged or securitized such that high-risk products (BB–CCC) were combined with low-risk products (AAA–BBB), enabling the securitized product to be A-rated (AAA, AA, or A). For example, while the home mortgages of individual consumers who are unemployed and without any substantial assets are rated CCC, once their mortgage is combined with many other higher-rated mortgages and securitized by an investment bank, that investment product may receive an AAA to BBB rating. As demand for investment products grew, and lenders ran out of low-risk mortgage products, they began creating securitized products with higher-risk borrowers. The investment grade that the securitized product received no longer reflected the individual mortgages within the package. The investors purchasing the securitized products from the investment banks no longer knew the grade of the individual mortgages they held or their own exposure to risk. Investment banks like Bear Stearns and Lehman Brothers were either unaware of, or concealed, their own risk exposure in an attempt to sell their products to investors.

Further, the investment banks began to borrow against their securitized products and lend money out for other investments. When a bank borrows money against its own capital, it is called leverage. Effectively, a bank leverages its capital to raise additional capital. Borrowing one dollar against one dollar in capital has a leverage ratio of 1:1. But banks frequently borrow against their capital at a much higher leverage rate than 1:1. When Lehman Brothers went bankrupt, its leverage ratio was 44:1.

As the economy began to deteriorate, individual borrowers who held the mortgages within the securitized products began to default because they could no longer pay the interest on their mortgages. As homeowners defaulted and houses went on the market to be sold, the housing market became depressed, which decreased housing prices and made it harder for all homeowners to sell. A rush to sell houses developed, as borrowers desperately tried to get their money out of the housing market before it collapsed. Greed was usurping any naturally correcting free-market control mechanism. A huge house of cards had been built on loans that would never be paid back, and that house of cards toppled in September 2008.

When Lehman Brothers went bankrupt, the entire system of banks around the world began to freeze as there was nothing any of them could do to mitigate their exposure. Lending stopped, consumer confidence plummeted, and a global recession began.

A week later, Ben Bernanke, the new chairman of the U.S. Federal Reserve, who had replaced Greenspan, convened an emergency meeting to avert further collapse of the remaining U.S. banks. Action was imperative, as the financial institutions involved were supposedly "too big to fail." The fear was that if they did fail, the entire global monetary system would crumble.

The consequence of globalization

According to John Gray, the "laws of economics" have been invoked to encourage one type of system, the "free market," which is meant to be the economic model for people everywhere. As we see from the earlier discussion, one of the consequences of the global free market is that it lacks checks and balances, making it susceptible to collapse. This is especially true when the system can be manipulated by individuals corrupted by greed. As Alan Greenspan said, "The question isn't whether or not competitive markets function perfectly, they do not. Regrettably there is nothing better."[11]

Another consequence of globalization is a reduction in diversity throughout the world. Dominant corporations and economies outcompete smaller, marginal entities. Although this may improve production efficiency, it also means there is less variety. There is a sameness in the world that did not exist fifty years ago. It may be comforting to depend on a brand-name hotel and restaurant – a familiar home away from home – during our travels, but in the process we lose something precious: the local uniqueness and variability that met and reflected local needs, creativity, and demands. Locally innovative ways of doing business provided options for varying environments.

By removing these options, the economic world has a more limited capacity and fewer novel ideas to help it adjust to a less stable environment. For example, we are losing the number of food options as monopolization and international corporations reduce variability by producing monocrops and genetically modified agricultural crops.

The impacts of globalization and the sameness associated with corporate culture and consumer goods are not limited to economies. Gray noted that "a nearly universal consensus proclaimed that globalization was forcing a movement to the centre ground of politics. In fact, as could be foreseen, it has fuelled extremism."[12] In Europe, the United States, and Alberta, Canada, we have seen a resurgence of the far right. This is no accident. As Gray wrote, "The radical right understands the fragility of liberal societies better than most of their defenders."[13]

Those of us in liberal democratic capitalist societies are encouraged to believe that capitalism – or perhaps more appropriately "corporate capitalism," as economist John Kenneth Galbraith termed it – is the same thing as democracy, and that democracy, defined by the *Oxford Canadian Dictionary* as "a classless and tolerant form of society," will cure all ills, even terrorism. Yet we confuse the democracy afforded to the market system with democracy for humanity. The capitalist economic system stimulates the production and distribution of goods. It depends on private capital investment and profit making. Positive and corrective competition is meant to generate continued growth of a healthy economy. The democracy of the market system is not intended to create equality for all humanity regardless of their capacity to contribute or consume.

This is particularly evident today. In 2003, John Kenneth Galbraith told an interviewer, "I never foresaw, and few of us foresaw ... the openness with which a policy for the rich is pursued in ... everything from taxation to foreign policy."[14] Galbraith described how many of corporate capitalism's best managers moved on to government, introducing corporate practices into government, particularly in the United States.[15] The result is a shift in government policies. Rather than seeking to establish equality and security for all citizens, an increasing number of policies support a small group of powerful and influential lobbyists and encourage governments and citizens to consume at any cost, irrespective of potential negative impacts on citizens or the state.

Our environment and economy are linked

Another factor influences the economy: the environment. Environmental crises, whether natural or human-induced, have a fundamental impact on our economy. The 2011 magnitude 9.0 Tōhoku earthquake in Japan triggered a massive tsunami, lingering aftershocks, and a nuclear disaster with rolling blackouts and radiation. Nearly 16,000 people were killed, 4,647 are missing, and more than 300,000 buildings, 2,126 roads, and 56 bridges were destroyed or damaged.[16] This combined natural and human disaster has had a profound effect on the Japanese people and on the Japanese and global economies. Personal consumption has plummeted in this country, with the world's third-largest economy, and this is a concern for policy makers as they seek funds to pay for disaster reconstruction.

The Japanese crisis is not a singular event. In January 2010, a much smaller-magnitude (7.0) earthquake struck a significantly less prepared Haiti, killing 316,000 people, injuring 300,000, and leaving 1.3 million homeless.[17] The estimated total cost of the disaster ranges between $8 billion and $14 billion. In July 2010, floods in Pakistan inundated huge swaths of the country, washing away crops, destroying villages, and causing an estimated 1,700 deaths and $9.7 billion in damage. Those floods returned with a vengeance in September 2011. On December 26, 2004 the third-largest earthquake in recorded history (9.1) hit the west coast of northern Sumatra, killing more than 225,000 people, displacing about 1.7 million, and causing tsunamis in 14 countries.[18] The cost of aid and reconstruction has been estimated at $7.5 billion. The eruption of Eyjafjallajokull in Iceland in April 2010 resulted in airspace closures, stranding passengers, interrupting global supply chains, and costing airlines an estimated $2 billion. The list goes on. Environmental crises disrupt economies and markets, and scientists predict that we will be experiencing more extreme climate events.

The sheer size of the human population is also influencing the environment and the economy. The acquisitory nature of the global market system, which drives us and our governments to consume more and more and to continuously grow the economy, is in turn bankrupting nations and "altering global biophysical systems and processes in ways that jeopardize both global ecological stability and geopolitical stability."[19] The widening gap between rich and poor, locally and internationally, creates further political instability. These destabilizing forces push the economic market out of its typical self-correcting business cycle and into the freefall of depression and beyond. When the market is unstable, we become fearful, less tolerant of change, and more reluctant or unable to see the connection between social, economic, and environmental change. The consumption bias of the government and marketing professionals makes it difficult for us to choose not to consume when, in the interests of political, economic, and environmental stability, we should be consuming less and more wisely.

A system in need of change

Our assumptions and behavior bind us to habitual past patterns, making us unable to face the stark realities of how our behavior is destabilizing all Earth's systems, including the economic system. We cling to these past behaviors and habits in the hope that they will solve our latest problem. We expect that this will just be another typical business cycle that will correct itself. Surely Adam Smith's invisible hand and the forces of competition, survival of the fittest, and natural selection will stabilize our economy and our Earth for us. If we just work hard and keep doing what we have always done, things will improve. But as Albert Einstein said, "We can't solve problems by using the same kind of thinking we used when we created them." We can no longer depend on future economic or environmental stability. Alan Greenspan admits that the competitive free market system will suffer another crisis, and subsequent ones. Each crisis will be different.

Liberal democratic nations that have been operating in the competitive free market are so accustomed to power, to dominating other nations, and to acquisition that it is difficult to imagine how they can alter their behaviour. As C. B. Macpherson asks, "How can they drop it when the whole structure of their society has come to depend on power-seeking, both individual and national, both economic and political?"[20]

Alternatives do exist. This system, which was built and which thrived during an interval of relative stability, needs a new perspective if it is to adjust quickly enough to a rapidly changing environment. Its structure and participants must be open to rigorous oversight. At the same time, they must remain flexible and open to variety and innovative ways of operating. It will be important for all sectors of society to adjust, including governments, businesses, NGOs, religious and community groups, educational institutions, and individuals.

Despite the success and influence of liberal democratic societies, about two-thirds of the world's countries currently use a different form of political system, ranging from autocracy, theocracy, fascism, and monarchy to communism and socialism. Many of these societies hold that "the requisite equality of human rights or human freedoms cannot be provided in a market society."[21] In other words, the liberal democratic market system does not meet their needs. According to Macpherson, some of these societies have their own concept of democracy, which is broader than the liberal democracy of capitalist market societies. For them, democracy is more than a system of government, more than the freedom to acquire and to choose – it is "an ideal of human equality."[22] Others do not. In the early twenty-first century, these alternative societies possess power they previously did not. They have succeeded in voicing their opinions about world affairs.

Those on the fringes of the affluent, dominant society show us an alternative way. There, without wealth, an unlimited opportunity to acquire means little. Individuals on the fringe must be nimble and creative to survive, and they rely on some form of community to keep them safe. Theirs is a precarious life. Yet when economies collapse, many of us end up on the fringes of our normal world, as was clearly evident during the depression of the 1930s and is increasingly evident during the current global financial crisis.

Western countries still hold a substantial balance of power, but even in the mid-1960s, Macpherson saw that, in time, this power balance would shift, with liberal capitalist countries' power declining. He believed that in the future, power and influence would depend on "moral advantage." By this he meant that Western countries would need to discard their acquisitive market behavior in favor of a more humanitarian perspective. By providing massive aid to poor nations, liberal democracies could sustain both their moral stature and their power. Yet today, even as massive aid is being given, more is being demanded as the world experiences an escalating number and degree of crises.

When he presented his lectures in 1965, Macpherson was concerned about destabilization associated with inequality and the consequence of warfare between the great political powers. He could not have foreseen the impact that our acquisitive behavior would have on the climate or on the global economy. Scientists now can see that our present behavior is causing rapid and significant climate change. Economists predict that our past behavior and its consequences will further destabilize our capitalist market system. These factors may yet shift morality to center stage, with the previously dominant powers relegated to back-row seats. Alternatively, as John Gray predicts, population expansion combined with greed, mounting competition for natural resources, and the prevalence of weapons of mass destruction portend a more destructive human future.

In 1930, the economist John Maynard Keynes felt that it would take about a hundred years before "we shall once more value ends above means and prefer the good to the useful."[23] With less than two decades to go before Keynes predicted we would discard our practice of promoting the accumulation of capital at all costs, we must make massive changes in our behavior if we are to "conserve the moral stature and the power of the liberal-democracies," as Macpherson put it.

Change is imperative because, as we will see in the next chapter, societies possess a life cycle similar to business cycles and biological cycles. Just as some organisms have developed and flourished through a period of dominance only to be destabilized by outside forces that allow others to prevail, just as there have been ups and downs in the market, so too have dominant societies experienced cycles from dominance to demise.

19

The Progress of Dominance

> None of the classical economists believed that mathematics should be the model for social science. For Adam Smith and Adam Ferguson, economics was grounded in history. It was bound up inextricably with the rise and decline of nations and the struggle for power between different social groups.
>
> *John Gray,* Al Qaeda and What It Means to Be Modern

> The most successful economies are those that have the flexibility and dynamism to cope with and embrace change.
>
> *Sir Nicholas Stern,* Stern Review: The Economics of Climate Change

Over the past several millennia, dominant human societies have flowered and declined, including the complex, advanced Maya culture; the Roman Empire, with its huge influence on the modern world; and the British Empire, with its vast conquered territories and global political power. When a society is successful, its citizens come to believe that their way of managing people, politics, communities, religion, culture, education, and economics is the right or best way, and sometimes the only way. Prolonged success can reduce the society's capacity to adjust to impending change.

Darwin's theory of natural selection and survival of the fittest serves to reaffirm this attitude. The dominant entity, whether it is a society, culture, or business, is by definition the fittest. This naturally leads to the conclusion that its laws, morals, principles, business practices, religion, language, and culture are best and that all actions, behaviors, or attitudes that ensure success are, by definition, justified and fitting.

Over time, the dominant society, culture, business, or individual becomes secure in its position of dominance and is increasingly hesitant to accept alternative rules, principles, and practices. It maintains its dominance by controlling its environment, restricting change, and preventing any instability, whether environmental, societal, organizational, political, or individual. The desire to maintain "control" leads to a progressive narrowing of acceptable behaviors, which results in reduced tolerance of alternative cultures, languages, values, religions, ideas, and perspectives.

The dominant society may believe it has been selected and that it is the fittest, but it fears threats from lesser countries. When it is attacked, as the United States was on September 11, 2001, it retaliates in order to punish the attacker and verify its dominance. An attitude of "us (good) versus them (evil)" emerges and proliferates. Defense spending is boosted, new preemptive strategies are developed, and individual freedoms are curtailed to enhance the security of the state. In the case of the United States after 9/11, political philosopher John Gray wrote, the U.S. government assumed that these efforts and their costs, including the loss of human lives, would make the world safe by maintaining the status quo and entrenching America's global hegemony.[1]

However, ten years later, we question whether the United States has the capacity to secure a victory. Its primacy depends on its continued economic and political ascendance and the willingness of the rest of the world to continue to accept its place, but the United States is now the world's largest importer of capital; this has reduced its capacity to influence foreign financial and economic policy. Investors are nervous about the future strength of the American dollar. Some, like the Saudis, who play a pivotal role in global financial markets, may be distrustful of, if not hostile to, American foreign policy. Mainland China continues to be the single biggest foreign holder of U.S. Treasury securities.[2] Large-scale withdrawal of capital would further destabilize the already fragile global economy and exacerbate the ongoing global financial crisis. Further, the ability of America to retain its place as the "sole military mega-power" depends on its economic predominance, which has been badly eroded by the recent global economic crises. Perhaps most critically, America has become extraordinarily dependent on imported oil to maintain its economy and to facilitate its military objectives.[3] Yet, stunningly, but perhaps not surprisingly, despite a new administration, the government remains gridlocked, unable to pass policies that would facilitate the development of alternative energy solutions to reduce the country's dependence on fossil fuels.

Unfortunately, the mentality of "good versus evil," in which the "fit" are good and the "unfit" evil, results in the destruction of life, culture, variability, and economies in war-torn societies; it fuels anger and distrust. It also results in the destruction of diversity and democracy at home as intolerance increases and difference becomes something to fear.

A reluctance to encourage diversity can also be seen in dominant businesses where, after a prolonged period of market success, corporate executives and managers lose their capacity to develop and cultivate new ideas within their firms. For example, large companies in the oil and gas and mining industries previously maintained extensive research departments and mandates to develop new properties and innovative technologies. Now many of them depend on small companies and prospectors to find opportunities. Only when significant resources are confirmed do they step in with investment and development funds.

The same attitude is evident in government, where elected officials and bureaucrats implement ideas and policies that reflect the needs of the dominant elite, naively or stubbornly locked in the belief that what is good for the corporate elite is good for everyone else – the country, the world. This mindset limits the creativity that might produce initiatives that would help those citizens who are lagging farther and farther behind. The failure to help these marginalized people causes human suffering, political instability, and desperation – societal attributes that further destabilize the economy.[4]

Conversely, survival in nondominant societies, cultures, or businesses often depends on flexibility and nimbleness. Recognizing the importance of disparate worldviews and cooperating with people who hold them are critical strengths. A society or organization that encourages and celebrates difference also fosters an attitude of respect for and by individuals, engendering sympathy and understanding of others' cultures, ideas, and perspectives. A culture that stimulates and encourages variety – whether it is in business, politics, or society at large – will, by its very nature, broaden the focus and perspective of its citizens and cultivate tolerance, if not compassion.

Thus, while a dominant attitude narrows the number of acceptable behaviors and limits variety, thereby decreasing potential available alternatives, a tolerant attitude stimulates variety and increases the number of potential options available to implement change. This is because a citizenry – whether societal, organizational, or individual – with multiple perspectives possesses greater intellectual and emotional freedom to generate, develop, and discuss new ideas and will more easily employ a varied set of alternatives during periods of change. Further, because such a citizenry is acclimatized to accept behavioral differences, adjustments in behavior are more likely to be tolerated and successfully implemented.

During an interval of economic or market stability, those societies, organizations, and individuals who have been dominant can maintain the status quo and can reasonably expect to continue to generate wealth directly for the corporate power brokers, the dominant organizations' shareholders, the country, and the citizenry. In contrast, tolerant and multidimensional societies, organizations, and individuals, although they possess enhanced intellectual, social, and cultural variety, do not have an opportunity to take full advantage of novel contributions.

However, during a period of instability, a country or corporation that clings to the status quo will confine itself to an intellectual and behavioral straitjacket that restricts its ability to adjust and erodes what was previously seen as a secure and unassailable position of dominance. A society or business that has nurtured a culture of flexibility and acceptance will not only have more viable options from which to choose, but will also be preconditioned to accept that there is more than one way to resolve a problem. It will be more able to adjust to change.

Today, significant change and instability are evident. There are more than 7 billion people on Earth, many of whom are feeling the repercussions of a global financial crisis and/or are being battered by extreme natural and climatic events. Demographers predict that global human population will surpass 9 billion by 2050. Darwinian natural selection helped us remain genetically unchanged as we managed this tremendous increase in population and held our position of dominance. We were also helped by a period of relatively stable climate that was made even more stable by behavioral and technological changes that humans implemented over the last 10,000 years. However, that dominance is now stimulating climate change unprecedented in human history, which will exacerbate economic, social, environmental, and cultural instability that is already beginning to appear. The practices of presently dominant human societies and corporations are unsustainable, especially as the global population continues to expand. *Homo sapiens* must remember that, like all other previously dominant species on Earth, we have not been granted *carte blanche* to be forever dominant – or even forever present.

In "the most comprehensive review ever carried out on the economics of climate change,"[5] Nicholas Stern, former chief economist of the World Bank and head of the United Kingdom's Government Economic Service, wrote that in the face of such change, "the most successful economies [will be] those that have the flexibility and dynamism to embrace the change."[6] Nimble, flexible companies, governments, and societies that can quickly change their behavior will have an advantage.[7]

As the second law of thermodynamics tells us, "any complex differentiated system has a natural tendency to erode, dissipate, and unravel."[8] It is no wonder then that the onset of rapid or significant change in the environment makes the market system and dominant organizations prone to disruption, disequilibration, and destruction. The result, whether we are prepared or not, is that change is upon us. The question is: Do we have the ability to accept responsibility and to change our behavior?

Dangerous Attitudes

20

Dangerous Attitudes

> We can't solve problems by using the same kind of thinking we used when we created them.
>
> *Albert Einstein*

Despite intense efforts by scientists around the world, climate change has remained, until recently, mainly a scientific issue. This is unfortunate, as it has cost valuable time and placed the public at a disadvantage, without a clear understanding of the situation. Poor communication of the issue, and the resulting noncommittal global response to climate change, are at least partly owing to the reluctance of those in positions of dominance, particularly those whose dominance is a result of the fossil fuel and auto industries, to accept or admit that dangerous change is upon us. This dominant elite, including government officials, continues to benefit from the status quo, while the media, which could be just as adept at convincing people of the imminence of climate change and its impact on our economic and physical well-being as it is at selling new cars, has frequently buried the issue under controversy over conflicting personal opinions while obscuring the scientific facts.

I recently submitted an abstract based on some of the material covered in this book to organizers of a conference predominantly for oil and gas professionals. One conference executive - a consultant to the oil and gas industry who was charged with reviewing my expanded abstract - encouraged me to reconsider my approach before making my oral presentation. I double-checked the validity of my statements, corrected some minor errors that he had highlighted, ignored the more stinging and unsubstantiated criticisms, and presented my research. My oral presentation was extremely well received.

Despite a preponderance of voices to the contrary, there is a willingness to listen, particularly in face-to-face encounters. If engaged action is to follow that listening, we must look at how and why we do things.

For instance, when I arrived at the airport en route to this conference, I discovered that although there were plenty of rental cars, there were no buses or taxis to take me to my hotel. I had to hire a limousine, which I shared with another colleague. The cost per person was the same whether we went together or hired the car individually.

The hotel was situated across the street from the conference center, an easy three-minute walk in a safe neighborhood, but buses left the front door of the hotel every fifteen minutes to take people to the conference center. Even so, I noticed that many people were not walking or taking the bus, but were climbing into their rental cars, driving across the street, and parking in the conference center's parking lot. I walked from the hotel, frequently arriving before those who drove.

At the center, I discovered that the only means of gaining access to the upstairs conference rooms was by escalator. There were stairs, but they were behind a locked door.

As I walked back to the hotel at the end of the day, the bus driver, seeing me standing on the corner of the crosswalk, stopped and asked if I needed a ride the 50 meters to the hotel's main entrance. I thanked her and walked, arriving before the bus. I met a couple of Europeans who, like me, walked and could not understand why so many people would not just walk across the street.

During one of the conference seminars, a group of us were discussing the capacity of modern humans to respond to changes in our environment, particularly our ability to migrate to new regions after catastrophic climate events. One of the fellows engaged in this discussion suggested that we were far more capable of adapting to climate change than our ancestors were. As we sat in our freezing-cold, air-conditioned meeting room on this warm late summer afternoon, I glanced across the table at his relatively young, hunched, overweight body slumped over his laptop computer and was perplexed. When the meeting concluded, he used his cell phone to call for a vehicle to take him to his hotel. He seemed to miss the irony that modern humans have become incredibly dependent on fossil fuel technology to sustain us and keep us warm (or cool), fed, connected, and mobile. Further, our physical capacity is declining as our dependence on technology increases and we create more and more mechanisms to perform our physical work for us. The aphorism "Use it or lose it" has never been more relevant.

Given this context, is it any wonder that it has been difficult to get the message of climate change out in a meaningful way that people can understand and relate to in their daily lives? It has even been hard for people working in climate science to understand the physical, cultural, economic, and political implications of climate change. So it was with interest that I learned, in late October 2006, of British foreign secretary Margaret Beckett's attempt to reconfigure the debate about climate change. Taking the stance that climate change is a global security issue, she argued that it has caused the growing instability and conflict in the Darfur region of Sudan. Climate change had reduced the productivity of rain-starved lands and was exacerbating conflict as competition for scarce productive land intensified.[1] The same is now being said of the famine in Somalia.

Although this is a welcome and novel perspective, it is imperative that we not rely on superficial messaging.[2] These crises are the product of decades of war and complex racial and political dynamics. Blaming everything on climate change simply reduces the legitimacy people give to valid climate concerns. The fact is that climate change aggravates such divisive conflicts; the repercussions will continue to grow if we insist on maintaining the status quo, refuse to change our attitudes and behaviors, and fail to come up with permanent solutions that deal with root causes.

One week after Beckett made her statement, Sir Nicholas Stern of the UK Treasury presented climate change from an economic perspective, addressing how it will affect the global economy. According to his *Stern Review on the Economics of Climate Change*, changes in the climate are expected to reduce living standards by 20 percent and plunge the world into a recession worse than the one in the 1930s. The economic cost, in this century alone, is estimated to be twenty times more than it would have cost to solve the climate change problem entirely. Stern's report caught the attention of politicians, economists, business people, and the public. Yet even before the predicted effects of climate change came to pass, the world plunged into a global financial recession.

Dr. Anwar A. Abdullah describes the real problem as a "clash between nations and nature.... We have imposed our will on nature through technology. Yet nature is silent, and humans speak. We are both the questions and the answer."[3]

> In fact, the merging breakdown between modern economic development and surrounding ecosystems could be reached back to the deep past of our civilizations. Since the early building of Mesopotamian city-walls, and of the French revolution, mankind has suffered from the illusion of nations; each against all other and all in all against nature. Here, nature is considered as stock of economic goods while the rising societies, of what has become known as nations of mighty urban-civilizations, have inflected their technological must upon silent flora and fauna. Paradoxical themes like these do reflect indeed the instability of cultural orientation, and of rather a permanent expression of mastery: of "man-and-nature" is little else but of "master-and-slave." And thus, to acquire a place on Earth, each nation has to justify its means of mastery over the rest; men and nature, whilst to feel right even when doing wrong. Here, the silent song of nature has never been subjected to a heartily detection. Nor its quality is praised beyond the claim of dry equations. So the current dilemma seems almost as a deep clash between nation and nature. And still, nature has its own stratagems as to absorb our blows and fool us whilst turning them upon mankind an utter wrath.[4]

The world needs novel perspectives. New leaders are standing up, risking much to be heard. They know that they might make mistakes, be ridiculed, and lose everything that used to matter to them. Yet it is precisely by taking these risks that positive change will occur.

As Einstein suggested, *new* thinking is required; if humans insist on maintaining our current controlling attitudes and behaviors, we will likely join Earth's previously dominant species, becoming fossil fuels for the next cohort of prevailing species.

Darwin was right about species evolving over long periods of time, but he believed species went extinct because they were not fit. In reality, fit species have become extinct throughout Earth's history because they were not able to evolve quickly enough when the environment changed rapidly. The result has been a changing parade of dominant and successful species over millennia. *Homo sapiens* has led the charge for the past 10,000 years, attempting to control Earth and further stabilize our environment, but although initially it appeared as though we were succeeding, the reality is that we have reached the lull before the storm. Earth is being pushed into another phase in a climate and economic cycle that will, if history repeats itself, require rapid change instead of gradual evolutionary adaptation.

We need to recognize that neither control of the environment nor slow, steady, minor adaptations to the status quo will solve our climate problems; neither will they solve our economic woes. Our attempts to control the environment are simply making matters worse. Significant and rapid change is required, and our capacity to accept difference, encourage variability, and develop tolerance will facilitate this change. It is the dominant attitude that needs to fossilize, not the entire human race.

In large part our difficulty stems from our belief that we are separate from nature, detached. We forget that we are just one part of nature, that we only presume to control nature, and that we rely on many other organisms for our survival.

21

Helpful Strangers

> The farther and more deeply we penetrate into matter, by means of increasingly powerful methods, the more we are confounded by the interdependence of its parts.
>
> *Pierre Teilhard de Chardin,* The Phenomenon of Man

> We are told that the trouble with Modern Man is that he has been trying to detach himself from nature.
>
> *Lewis Thomas,* The Lives of a Cell

Yet detachment is an illusion. As Lewis Thomas wrote in *The Lives of a Cell*, "Man is embedded in nature.... The new, hard problem will be to cope with the dawning, intensifying realization of just how interlocked we are."[1] He went on to point out that mitochondria, the energy sources found within each of our cells, are actually separate miniature creatures. They probably originated as primitive bacteria that found their way into our early ancestor's eukaryotic cells and subsequently resided there. Mitochondria possess their own DNA and RNA (ribonucleic acid), different from ours, and have their own means of replicating. They are just one of numerous examples of microscopic organisms, like bacteria, that reside within our bodies, depending on us and sometimes, like the bacteria in our intestines that aid in our digestion, providing the mechanisms for our survival. Without our symbiotic mitochondria, we would lack the energy required to think, move, act, or even exist. They have remained small, continuing to reside within us in a secure, relatively risk-free environment. The result has been a mutually beneficial, symbiotic relationship.

Knowledge of mitochondria leads me to reflect on my own identity. Who am I? I have always thought of myself as one entity, but am I actually a conglomeration of entities housed within one body, my existence dependent on theirs, theirs dependent on mine? Should I be more careful about what I eat, how I treat my body? Does what I do help or hinder these strangers residing in my body? Even before I knew they were there, my symbiotic aliens were busy helping me survive – helpful strangers. The self-same strangers that reside in you are also in your brother, sister, spouse, mother, father, uncle, aunt, nephew, niece, neighbor, friend, stranger, dog, cat, goat, cow, seagull, bear, beetle, crow, fish, blade of grass – helping them create energy, breathe, grow, and die. They emphasize our interdependence and our connectedness.

In 1974, when Thomas wrote *The Lives of a Cell*, fewer than 4 billion people lived on Earth. Thirty years later, in 2004, Earth's human population exceeded 6 billion; by 2011, it reached 7 billion. If, as Thomas wrote in *The Lives of a Cell*, "everything that comes alive seems to be in trade for something that dies, cell for cell,"[2] many individuals from different species must have given up their lives and not been subsequently replaced to fuel *Homo sapiens*' expansion. The human population is expected to balloon to more than 9 billion by 2050. If this happens, what shall have traded their lives, "cell for cell," for the next human generation? Can Earth and its other species sustain this expansion? Or will we have – or have we already – reached its tipping point, the point beyond which humans consume more than Earth's remaining species can provide?

In 1996, William E. Rees, an ecological economist and professor at the University of British Columbia's School of Community and Regional Planning, wrote that "the fundamental ecological question for sustainability is whether remaining *natural* capital stocks (including other species populations and ecosystems) are adequate to provide the resources consumed, and assimilate the wastes produced by the anticipated human population into the next century, while simultaneously maintaining the general life support functions of the ecosphere."[3] Fifteen years later, both population and consumption continue to escalate while productive land and resource stocks steadily decline. Developed nations talk about falling reproduction rates as a "problem" because this diminishes the workforce and market potential necessary for an expanding economy. Expanding populations are considered inevitable. Yet humanity has been consuming far beyond Earth's capacity to support this species for a long time.

In the November 3, 2006 issue of *Science* magazine, international researchers declared that declining biodiversity is causing large-scale global reductions in the size and robustness of local wild fish stocks.

Although the study did not attribute the reduction of biodiversity to pollution, overfishing, or habitat loss, it did imply that marine protected areas generate greater marine biodiversity and more and larger fish. If the trend of reduced fish stocks continues, within fifty years there will be virtually nothing left to fish. As discussed in Chapter 17, there is also fear that increasing ocean acidification will lead to the collapse of coral reefs, causing "serious consequences for reef-associated fisheries, tourism, coastal protection, and people."[4]

These concerns have stimulated research to better understand our marine life. The oceans remain an unknown frontier. On October 4, 2010, the "Census of Marine Life" published results from a ten-year international study assessing the diversity, distribution, and number of different types of marine life around the world.[5] What emerged was a picture of ocean life. The researchers found 1,200 previously unknown marine species, with 5,000 more yet to be described. They discovered giant mats of microbes that are some of the largest masses of life on Earth. "They unearthed a rare biosphere in the microbial world, where scarce species lie in wait to become dominant if change goes their way, and found species believed to reside at both poles." They documented "long-term and widespread declines in marine life as well as resilience of the ocean in areas where recovery was apparent." They determined that coastal and closed seas have been most affected by human activity, and predicted that the greatest impact on marine life in the future will come from climate change.

The maintenance of marine biodiversity is critically important because, as Dr. Boris Worm of Dalhousie University states, "all parts of [marine life] are integral to the structure; if you remove parts, particularly at the bottom, it's detrimental to everything on top and threatens the whole structure."[6] Scientists fear that the complete and terminal decline of cod that resulted from overfishing along the Grand Banks off Canada's east coast could be replicated in fish stocks worldwide if politicians continue to disregard recommendations for marine protection. There are earlier examples that are relevant and that have been similarly ignored, including the collapse of the North Sea herring fishery over a mere twenty years between 1960 and 1980.[7]

Throughout Earth's long history, it appears that climate catastrophes have been responsible for mass extinctions. Such climatic crises create a condition in which insufficient resources are available to support and sustain the dominant species. Today, we are perilously close to the same circumstance, except this time the impetus for climate change is human-induced.

However, even in the absence of climate change, our "right to reproduce" threatens to destroy every other human right. Our desire to maintain our dominant status, our commitment to survival of the fittest and the corporate market system, our habits of overconsumption, our insistence on detaching ourselves from nature, and our attempts to control Earth have led us to the edge of the abyss that we have been so desperate to avoid.

Homo sapiens now presides as the dominant species on Earth. We are triumphant, yet we remain oblivious to our symbiotic and dependent nature. We are dependent on other organisms for our very survival. Whether those organisms reside within our bodies, provide the oxygen that we breathe, or give up their lives to provide the food we eat, there is no doubt that we rely on them. Without them we perish. Perhaps it is because we are so desperately dependent on Earth and its creatures that we choose to remain blissfully ignorant of these helpful strangers. The magnitude of our dependence makes us frighteningly fragile and reliant, and to accept our dependence would require that we also admit our weakness, an oxymoron for the "fittest" species.

Recognizing our symbiotic and dependent nature is the first step in healing ourselves and our world. It is also the first step in recognizing that our survival depends on identifying a new perspective, one in which survival is not solely the privilege, or the consequence, of being the most dominant.

22

Triumphant Oblivion

> If we conceive of culture as one body, which it is, we see that all of its disciplines are everybody's business.
>
> *Wendell Berry,* The Unsettling of America

Within the last decade, scientists have discovered evidence of what appears to be water on Mars, and NASA was recently planning to put people on that planet. Scientists and politicians justify this multi-billion-dollar project with talk of military objectives and climate change. I cannot help but cringe. We have not been able to accept responsibility for our impacts on this planet, yet we are going to impose ourselves, with our coincident blunders, on another. My hope is that, by the time we get to Mars, we will be aware of our blind ineptitude and will have made concerted efforts to change our perspective and our behavior.

According to our cherished paradigm of natural selection, with its survival of the fittest and slow, steady, minor variations, life should always be evolving to a better state, yet we appear to be in serious danger of destroying the environment that sustains us. If our species and others are in danger, and if natural selection, survival of the fittest, and the corporate market system that reflects this paradigm will not lead us out of the mess, what will?

We need a new perspective. We need to recognize that we exist in a symbiotic and dependent relationship with ourselves, other species, and the Earth, and that our survival is not solely the privilege or consequence of being the most dominant, the species that can destroy others weaker than ourselves. If destruction of others results in our own destruction, we are not the "fittest" or most dominant species; we are simply the most accountable species on Earth, on whose shoulders rests the obligation to leave our world better than we found it.

A new perspective will allow us to embrace and participate in change, not shun it. Our tattered paradigm of natural selection favors slow, gradual, genetic change over millennia. It encourages us to disassociate ourselves from responsibility for change by placing the capacity for change solely within our genes. We raise our hands in surrender, yielding to the self-fulfilling prophecy that what happens must be right because only the fit survive.

Yet historical evidence indicates that new species can emerge rapidly after a catastrophic event, which is a good thing. Otherwise, each catastrophic change in Earth's climate would have further reduced an ever-depleting array of species. Any new perspective must recognize this capacity for rapid change and accept that we not only have the capacity to change quickly, but also have a responsibility to do so when it becomes clear that we must.

It appears that polar bears have already recognized that a crisis exists. In the spring of 2010, a couple hunting on Victoria Island in the Arctic were tracking a bear that had been destroying hunting cabins along the frozen coastline. They caught up with this bear as it crossed the open plain, and they shot it. But there was something very odd about its wide head, brown paws, and matted white fur, so after they took it home they had the government wildlife officer sample it for DNA. The DNA indicated it was a grizzly–polar bear cross, meaning it was the offspring of a grizzly bear and a polar bear that had mated. They also discovered that this hybrid bear's mother was a grizzly–polar bear cross too.[1] This means that not only did two separate species mate, but they produced a fertile offspring – contradicting the biological definition of species, which is organisms that can breed and produce fertile offspring. The hunting couple had discovered a new "hopeful monster."

The environment is the milieu in which change occurs; it also influences that change. In his book *Biological Emergences: Evolution by Natural Experiment*, Robert Reid shows how such emergent change can occur at the level of developmental epigenetics, as when individuals inherit characteristics caused by adverse events in the environment, even though they did not actually experience those events themselves (as described in Chapter 15).[2] Such change can also happen through physiological adjustments, illustrated by the salamanders in Paul Kammerer's lab or the pygmies in Africa. But perhaps most importantly for humans, change can happen as a result of the symbiosis of societies. When disparate societies, cultures, and perspectives come together, it is a boon to human behavioral adaptability. The conglomeration of perspectives, backgrounds, and expertise generates new ideas – technological, scientific, cultural, religious, and educational. The outcome is astonishing if we allow ourselves to go there; the result is greater than the sum of the parts.

Unfortunately, our selectionist paradigm has driven us to extreme reductionism, narrowing our focus to ever-smaller minutiae. This has been interesting and informative, particularly on the genetic front, and is consistent with natural selection's spotlight on genes as the determinants of change. However, we have devoted ourselves so completely to reductionism that we have done little to assemble minutiae into a bigger picture. In our universities, knowledge is dismembered into disciplines, which are further dissected into specialties; students know more and more about less. We have become unconnected and triumphantly oblivious to the events outside our miniature worlds. Generalists are vanishing. Our research institutions, universities, and funding agencies are not designed to support basic, general, problem-based research instead of, or in addition to, discipline-based research. Yet an interdisciplinary perspective is what we need to address the complex issues we are facing.

We need a new way of looking at the human condition, a perspective that will limit our triumphant oblivion, encourage our capacity to adjust to changing circumstances, and engender confidence that we are not only capable of changing but obligated to change – quickly.

That change is already occurring. For instance, new research suggests that not all bacteria that grow in the dirt are bad; researchers have discovered how to use harmless soil bacteria to help kill cancer tumors.[3] More helpful strangers! Bacteria are also being used to clean soil contaminated with fossil fuels. Within days, the bacteria produce clean, nutrient-rich topsoil in which plants thrive. But bacteria are also being used to create biological weapons of mass destruction.

So, yes, we need to be nimble and flexible. We need to act. We need to recognize that our short-term perspective is detrimental to the environment and to our children. But encouraging change alone will not suffice. Our new perspective must also allow us to recognize the plight of others – other humans and other species. It must embrace many disciplines and facilitate an awareness of difference, allowing us to bask in its benefits. Our new perspective must also encourage variability and nurture tolerance.

By turning to this new perspective, we will be able to develop novel ideas, behaviors, and conditions and improve our capacity to be tolerant and flexible, to embrace change. Our new perspective will apply equally to all races, colors, religions, cultures, languages, ages, and sexes. It will recognize that leadership is not dominance, that dominance leads to destruction, and that destruction is absolute. By recognizing our connection to other species and our planet, we will generate hope for ourselves, other species, and our environment.

What part in this new perspective is ours to play?

Living in Dangerous Times

23

Our Children

Blessed are the flexible, for they shall not be bent out of shape.

Anonymous

Newborns use their tiny hands to grasp their feet and touch their toes to their nose. They observe us with spongelike minds and emulate our every nuance. As they grow up they gain experience and knowledge, which by adulthood frequently develop into habits and rigid opinions. Although these habitual ways of being, thinking, and doing sometimes make it easier to deal with life's immediate challenges, their continued use makes us rigid. We lose our objectivity and capacity to change.

All good parents wish happiness, good health, and success for their children. We want them to participate in relationships and endeavors that will allow them to become the best they can be. We wish for them to have true wisdom and compassion and to avoid the mistakes that we have made. We want them to be happy. We tend to forget that children naturally possess a capacity and willingness to accept difference and change that helps them thrive in a rapidly changing mind, body, and environment. Perhaps this is why children are so flexible when they are born. It is only with age that we become rigid.

Rigidity encourages reluctance to acceptance difference; it stimulates inequality, or at least allows us to accept it. The consequences will not benefit our children.

Today, inequality between the "haves" and the "have nots," ethnic and religious differences, the scarcity of natural resources, and the collision of rival values are stimulating political unrest. While some parents strive desperately to feed their children, others argue over which SUV they will purchase as their third vehicle. The "haves" are consuming far more than their grandparents, parents, and others around the world have done. It is becoming increasingly evident that we are also consuming resources that our own children, as well as all other children, will need in the future.

The old paradigm of natural selection and survival of the fittest presents as reality the view that some will suffer at the expense of others. Supporters of that paradigm are thus justified in, regretfully, failing to act, because nature has clearly shown that some humans are more "fit" than others who are less worthy - and therefore expendable.

The alternative perspective, put forward in this book, recognizes that those on the margins, those less successful both inside and outside the dominant stable societies - the un- or less fit - are equally our children, and they possess a critical capacity and opportunity for change and, thus, hope. These are the individuals who, because of difficult circumstances, have been, or will be, forced to recognize limits to their needs and to survive under conditions imposed on them. These individuals, whether it is their desire or not, must retain or develop their capacity to be flexible.

This is quite different from the circumstances enjoyed by those of us who have failed to define any limits to our needs and who live under controlled conditions that create the most favorable environment for our limited survival. However, it is just these controlled conditions that are most likely to be severely disrupted during future climate and economic changes. Under these circumstances, "fitness" and "success" could mean great wealth and influence, or, ultimately, great destruction.

In temperate Victoria, Canada, a heavy snowfall is such an unusual event that it often results in children being given a day off from school. During the winter of 2006–7, a series of unusual storms hit the city, and the temperature dropped below zero for several days. The strong winds accompanying the storms caused prolonged power outages. A few individuals made some unfortunate decisions or were unable to adjust their behavior or circumstances in response to this relatively infrequent and minor environmental perturbation. Despite the storms' relative brevity, they died. It was clear that a capacity to remain informed and flexible allowed some to adjust to the rapidly changing environment better than others.

In Africa, developing a capacity to adjust to changed circumstances has sometimes meant that an eight-year-old child becomes the head of the household in a family devastated by AIDS.[1] It has meant that children with no one to support them and mentor them may steal for survival. This is another form of flexibility, but it foreshadows a world in which human battles human and we are all out for ourselves and our family alone, with no care for anyone else.

If we wish to improve humanity's future and move beyond our state of dominance, ignorance, and denial, there are two things we need to face. The first is that our actions can negatively impact others – and that when they do, it is imperative that we clearly understand how we are contributing to that hurt and determine methods to prevent such consequences. For example, we need to consider the impact of our fossil fuel use on climate change and on reductions in edible resources and fresh water around the world. We need to think beyond the short-term results of our choices to the long-term effects on our children, other species, and the planet. These problems will not go away until we take responsibility for our contribution to their existence. The world's children are the victims, not the cause.

Second, if we continue to marginalize people who have different habits, religions, skin colors, sexes, ages, cultures, education, economic status, and so forth, we will only wreak additional harm on humankind. Humanity depends on the great diversity of opinions, perspectives, and strategies in order to successfully deal with future change.

Even if we are not able to move beyond our dominant, selfish perspective in this generation, perhaps we can at least understand our interconnection with these people we see as different. Many of the people marginalized by developed societies are the same people who grow our food, sew our clothes, build our technologically advanced tools, provide us with our natural resources, and purchase our products. Further, not only are we dependent on these individuals for our success, but we are also dependent on a multitude of other species on the planet for our very survival. If a dominant society's behavior continues to negatively impact other societies and species on which it depends, the dominant society can no longer prevail.

You may know the story of Easter Island, a small island in the Pacific Ocean. The people who lived there hundreds of years ago chopped down the island's trees for cultural purposes and erected large stone statues. They continued chopping trees until every single one was gone. This so altered the environment of the island that it could no longer generate sufficient resources to support the people. What appears to have been a very advanced early society went extinct.

When the first astronauts went to the moon, they looked back through the spaceship window and were astounded by the beauty of

Earth and also by its fragility. If we think of Earth as an island, much like Easter Island, we will no doubt realize that its capacity is not limitless. Just as the Easter Islanders discovered, we too must understand that resources are finite.

In a changing world, a healthy, successful humanity will require a pool of diverse opinions, ideas, and cultures from which to draw future options. We will require a populace that is able and willing to change its behavior, one that is cognizant of Earth's capacity to support us. Those who have been placed on the periphery, who have been marginalized, or who are tolerant of difference have already been faced with a requirement to adjust to imposed or diverse conditions. They may have already begun to develop their own ideas and capacities for change.

A successful humanity will also need an educated and creative populace. Investment in education and basic research will be imperative – education that is not constrained by disciplinary, religious, economic, or other self-imposed boundaries, and research that delves into the unknown and the frightening, and pokes and prods established paradigms. It is by working together as individuals, societies, and cultures that we may make new discoveries, generate the necessary leadership for future change, and concern ourselves with the future of all the world's children. We can develop the capacity to feel what others feel, place ourselves into the life circumstances of other individuals, and understand the world as others see it. This does not mean that we will always agree with each other. It does not mean that we will be comfortable – at times we likely will be very uncomfortable – but there is a distinct possibility that a diverse perspective can improve our capacity to accept the worldviews of others and to feel the impacts of our behavior.

It is with the world's children that the future of humanity resides. They need to be given every chance for good health – physically, emotionally, intellectually, and spiritually – and their survival should not come at the cost of pitting one against another. For the future of our children, what will matter more than anything else will be how each individual lives his or her life, particularly as it relates to our overwhelming dominance on Earth. Each of us will have individual choices to make. We will be required to choose between what we need versus what we want, between our current existence and our children's future.

The corporate elite has achieved incredible wealth and influence, granted by the grace of consumers, nonconsumers on the periphery, future generations, other species, and Earth itself. It is on the shoulders of this elite that the greatest responsibility for change and leadership rests. It is to this group that our children look for future hope and leadership. But equally responsible will be those who have chosen to support causes over solutions, as it is always easier to speak than to act. Continued pitting of these two groups against each other will only undermine the future of our children.

As individuals, we must decide whether we wish to continue on a path destined to destroy the very things on which we depend for our success or whether we wish to choose a different future. Destruction will not be restricted to the nondominant. In fact, it will likely be equally, if not more, devastating to those who are most dominant and thus least willing to change.

We need to acknowledge the interdependence of religion, culture, ethnicities, linguistics, and disciplines. If we can develop the capacity to connect with our fellow human beings through these differences, perhaps we can then move beyond ourselves and begin to develop compassion for all the world's children and for the multitude of species on this planet on which we depend for survival. We need to share our knowledge so that as a species we can progress in a symbiotic way with all our fellow humans – helpful strangers or not – and all of Earth's other creatures. By doing so we will generate a healthier Earth and a healthier *Homo sapiens* for our children and their children.

24

Living in a Dangerous Climate

> With man gone will there be hope for gorilla?
> With gorilla gone will there be hope for man?
>
> *Daniel Quinn,* Ishmael: An Adventure of the Mind and Spirit

There is no question that these are challenging times. Yet humanity lived through a dangerous climate in earlier times. How can we live now?

Even if we work together globally to reduce population growth, the status quo is not an option. This is because we will not solve the climate change problem or our economic woes by behaving in the same manner that created the problem – by depending on the premise that the world was made for humans and humans were meant to conquer and rule it. This story casts humans as the enemy of the world. Unlike any other species on the planet, we compete wantonly with all other species: we exterminate them, systematically destroy their food supply, and restrict their access to food to make room for our own.[1] Ever since we developed agriculture some 10,000 years ago, we have maintained an inherent belief that control over nature is not only possible but beneficial. We uphold attitudes, based on the "laws" of natural selection and survival of the fittest, which presume that dominance equals fittest or best and that therefore, as the dominant species, whatever humans do is proper.

To suggest that bigger is always better and that competition and natural selection will ensure the survival of the dominant is no longer a tenable stance. History provides repeated evidence that, at least during periods of rapid change, such logic does not apply. Humans cannot control everything, particularly not Earth. We have been making a tremendous mess of things.

Darwin's theories of survival of the fittest and natural selection describe evolution during intervals of stable climate. When the environment is changing rapidly, these theories tend to break down. Rapid change requires and potentially stimulates saltatory, not slow and gradual, evolutionary development. And despite the fact that we believe that speciation stopped with humans, that we are the climax of evolution, speciation continues, as the story of the grizzly–polar bear hybrid exemplifies. This is because diversity is the factor that ensures the survival of life on Earth. It always has been. Earth cares not which species rules the day, only that life continues.

As I intimated in Chapter 1, some consider bacteria to be the dominant species on Earth, and they may yet prove themselves to be. Bacteria are incredibly flexible, capable of altering and taking advantage of changes in their environment. Humans are not so flexible. With the rapid climate change anticipated, it will be this capacity to be nimble, flexible, and able to adjust to a changing world that will allow a species to survive.

All of our cousin *Homo* species are now extinct, and despite the dominant paradigm's theories to the contrary, hopeful monsters have existed and continue to exist. Those that have survived have found success despite, not because of, natural selection and survival of the fittest. My "unfit" father survived by keeping fit, and he procreated. In the end, even fitness did not save him, yet his "unfit" genes reside in his offspring, where his memory also dwells. Woody Guthrie survived despite being "unfit", and he too had children, and his music continues to inspire us today. Pygmies survive and consider the rest of us unfit monsters. These "hopeful monsters" made, and others continue to make, their contributions to society despite their relative "unfitness." Others, like the "hopeful hobbits" and the Neanderthals, no longer do.

Adding to the neo-Darwinistic conundrum, we now know that epigenetic alterations caused by rapid environmental changes, like those of the Dutch Hunger Winter babies, can and do get passed on to offspring. Rapid physiological change can occur and be useful, despite Darwinistic theories to the contrary. This suggests that the ideas put forward by Kammerer and Goldschmidt and others, which imply that natural selection and survival of the fittest are not all-encompassing theories, have some substance.

Yet we continue to cling to the flawed paradigm, using it to justify our behavior, which may well result in the destruction of the delicately balanced system on which we all depend. We need to revisit our uncompromising belief that Darwin's theories hold under all circumstances. Natural selection is a hypothesis, not a law, and it does not always live up to its theoretical underpinnings. Our desire to consume and control everything and everyone but ourselves has already placed our species in a precarious state – hardly an example of the best and fittest surviving. Further, our denial of death – death of our habitual attachments, death of ourselves, and death of our species – has led us into a life driven by fear. None of us is infallible, even those "fittest" among us. Our conception of the world, our understanding of life, and our political, economic, and education systems are based on a flawed paradigm that describes not the origin of life but the restriction or even the destruction of life. This is evident in the pervasive military-industrial complex that makes a sham of the ideals of capitalism and democracy. It is no wonder that we find ourselves at this crossroads.

The good news is that when we allow ourselves to reconsider our uncompromising belief in the absolute universality of Darwin's theories, we discover hope. Specifically, we see that there is an opportunity to generate a rapid response to our self-inflicted state of chaos. We can choose change. We do not have to throw up our hands, reluctantly shift responsibility to our genes, and admit defeat because of their glacially slow, uncontrollable, naturally selected response. Instead, we can generate a positive, conscious response, and we have both the capacity and the responsibility to do so. Those businesses, societies, cultures, and individuals that have felt secure in their position of successful dominance will give way or evolve into tolerant, open-minded, nimble entities and individuals that have developed the ability to respect difference and the capacity to accept and adjust to change.

A new economic model is required. Consumers must believe in the viability and strength of their market and financial institutions for those institutions to persist. For that to happen, checks and balances must be put in place to ensure that the "free market" is not manipulated by individuals who are corrupted by greed. Diversity must be maintained, developed, and encouraged. We must become more open and accepting of variety. The benefits of corporate capitalism, including the generation of wealth and jobs, can come without insisting on universal sameness, the trouncing of diversity, and the destruction of all that is "other."

I live on the west coast of Canada, where some communities refuse to accept industrialism within their borders. They are "going green," but because they live some distance from city centers, they must drive to town to pick up their groceries and consumables. After shopping, they often meet with friends and neighbors at local coffee shops and have thoughtful discussions about climate change, the financial crisis, child poverty, and other important global issues. Most live in beautiful homes, set on rural treed lots and acreages. They try to buy local, but despite their best efforts, they cannot survive on their own. Much of their food and their computers, cars, appliances, and the like are grown or built in distant communities. Even the coffee they drink cannot be roasted and packaged locally, because people do not want the industry in their backyard. Other cities and towns benefit from producing the goods these west coast dwellers consume, but those communities also suffer the costs, including pollution and environmental degradation. Those who insist that their community not produce the products they consume, cut the trees that are used to build their homes, or mine the minerals and fossil fuels that are used to build and run their cars and computers simply put the burden on others' shoulders. Environmentalists who blame industrialists for all that is wrong will not solve the problem anymore than industrialists who blame environmentalists. Being part of a cause does not solve the problem; being part of the solution will.

We need to look at the world in a different way, one that recognizes we are in a rapidly changing world and that to survive and thrive we need to innovate and change.

What is innovation?

For humans, innovation is the capacity to develop new ideas, alter established behavior, and implement those new ideas and ways of being.

Why is innovation important?

Looking back on Earth's history, we have seen that when a crisis strikes, there are three options facing species: (1) move out of the affected areas; (2) innovate and change; or (3) go extinct. Humans are now the dominant species, but our options to respond to change remain the same. With 7 billion people on the planet and billions more expected to arrive over the next fifty years, we *cannot* migrate to somewhere new. We *can* innovate and change, or we can go extinct.

How innovative has the *Homo* species been?

Innovation at the species level creates variety, and there has been much variety in our past – *H. habilis, H. erectus, H. ergaster, H. heidelbergensis, H. neanderthalensis, H. floresiensis*. But today, there is only one remaining *Homo* species – *Homo sapiens*.

Why do we remain alone?

We remain alone because none of the other forms of *Homo* had the capacity to innovate and change their structure, physiology, behavior, or culture in order to adjust and survive in a changing world.

What made *H. sapiens* innovate and survive?

H. sapiens survived during previous rapid climate changes because of the three C's: crisis, communication, and collaboration. When crises hit, humans moved into restricted territories where they could survive. They brought with them different ways, responses, cultures, and behaviors. They communicated these different ways of being with each other. Then they collaborated. Intelligence emerged, as did innovative ideas and behaviors like complex stone tools, agriculture, and civilization.

How did we become the most dominant of all species on the planet?

Around 10,000 years ago, humans began to control and exploit plants, other animals, and nature generally. We responded to crises by increasingly controlling our environment so as to limit the amount of change with which we had to deal. We grew food, irrigated crops, stored food, heated and cooled our homes. We proliferated. Our dominance continued because our innovations kept up with the relatively minor climate and environmental changes we experienced. However, when innovations and behavioral adjustments did not keep up with a rapidly changing environment, extinction reared it dreaded head as the demise of Maya and Easter Islander civilizations attest.

What is wrong with dominance?

The problem with dominance is that it has protected us from change so we do not feel change or believe it can happen to us. Humans have become more and more specialized and less and less flexible. Without the push of crisis, we are less and less willing and able to change. Dominance has also allowed our population to grow enormously. We are consuming more and more of the planet's limited resources. Other species on which we depend for our survival are going extinct. Dominance has insulated us from the consequences of our actions. We are losing the ability to connect with Earth and understand its natural rhythms and cycles.

Why do we resist change and innovation?

In the past, humans changed not because they wanted to, but because they had to, because crisis forced them to change. Possibility was born from necessity. Today we resist change and innovation, in large measure, because we have yet to feel the crisis and therefore the need to change. Our tenuous connection with Earth makes us slow to recognize the necessity for change and respond. But we also resist change because our economic system is based on the theory of natural selection, competition, and survival of the fittest. That model tells us to focus on our immediate self-interests. It tells us that, over time, slow and gradual change creates the fittest species and economies. The problem is that model works in a stable environment when gradual change and evolution has time to respond. It does not work in a rapidly changing environment where slow, gradual adaptation does not happen fast enough to adjust to rapid disruptions in our economies, markets, and environment.

Our environment and economy are linked

We live in an interconnected system where isolation is impossible. We not only depend heavily on others for our survival, but environmental crises also have a fundamental and immediate impact on our economies – think here of Hurricane Katrina; the Tōhoku earthquake in Japan; floods in Pakistan, Haiti, and Australia; and droughts, food shortages, price hikes, and political unrest in North Africa. There are economic consequences to environmental crises whether they are natural or human-induced. Unstable environments disrupt economies and markets and exacerbate political conflict. And scientists predict we will be experiencing more extreme and destabilizing events.

Has Earth's climate always changed?

Climate change is not new. The Earth has experienced many extremes in climate that have taken it from snow- and ice-covered land- and seascapes to warm intervals when palm trees grew in the Arctic. Unfortunately, this has led some to suggest that we need not be concerned about the current changes. These people have missed the point.

What is different about today?

Climate change is not new, nor are species extinctions. What is new is the fact that the level of carbon dioxide in our atmosphere has escalated to levels never before experienced by *H. sapiens*, or observed in the scientific records that stretch over the last 800,000 years, and this has occurred as a direct consequence of human behavior. Over the past 160 years, the amount of carbon dioxide in the atmosphere has increased by the same amount it increased over the previous 21,000 years, a period during which the Earth moved out of a glacial deep freeze and into the moderate climate of the 1800s. Yet although we are able to predict that this latest increase in atmospheric carbon dioxide will lead to future climate change, we are unable to feel its full effects here and now because it takes time for these rapid atmospheric changes to work their way through Earth's climate system. What is different is that our current behavior will have long-term impacts on humanity and on all species on Earth. So although we can predict, we cannot yet feel the crisis – so little change is stimulated.

We need a new paradigm – the three C's

We need a paradigm that works in a rapidly changing world, a paradigm that encourages and stimulates innovation and change. And what stimulates innovation and change? The three C's: crisis, communication, and collaboration.

How are we doing today with the three C's? Crisis – some are already feeling it. Communication – innovative technologies are increasing communication between diverse people, but the question remains: Are we engaging in real dialogue? Collaboration – well … we could really use some work here. It is resisted by many groups, religions, cultures, and governments, and this threatens the capacity of humans to initiate change.

What do we need to succeed?

We need to recognize that Earth will persist with or without us. If *Homo sapiens* want to also persist, we need to accept responsibility for our actions. We need to recognize our dependence on each other, on other species, and on Earth. We need to accept and embrace diversity and difference. By doing so we will develop the wellspring that will feed innovation and the openness to implement change. We need to act.

Where will hope, leadership, and innovation come from?

Leadership will come from those individuals, institutions, and countries that are willing to collaborate with those of diverse languages, cultures, and behaviors; those places where diversity is embraced; and those institutions that encourage innovation and collaboration. Leadership will come from you and me.

Hope resides in a new economic perspective, one that does not involve solely maximizing the productivity of the economy, but rather adopting a more moderate economic approach that includes environmental and social considerations.

In 2007, the UN Environment Programme's Global Environment Outlook 4 released a study showing the implications of four different economic scenarios based on different global and regional policy approaches and societal choices.[2] In the "Markets First" scenario, the private sector, with active government support, pursues maximum economic growth and relies on technological fixes for environmental challenges. In the "Policy First" scenario, the government, with civil and private support, initiates strong environmental and human well-being policies while still emphasizing economic development. Competition between the private sector and the government in the "Security First" scenario focuses efforts on maintaining the well-being of the wealthy and powerful in society with limited regard for environmental sustainability. Finally, government, civil society, and the private sector work together in the "Sustainability First" scenario to increase human well-being, equity, and environmental protection. In this scenario, it is understood that results will take time.

Many economists and politicians from the developed world may find the results surprising. For me, they inspire hope. Under "Markets First," 13 percent of species go extinct by 2050, compared with 8 percent under "Sustainability First." "Markets First" generates 560 ppm atmospheric CO_2 by 2050, compared to 475 ppm for "Sustainability First." Interestingly, however, investment in social and environmental sustainability does not hinder economic development. GDP per capita in nearly all less-developed regions is higher in the "Sustainability First" and "Policy First" scenarios than in "Markets First" and "Security

First." There can be a greater investment in health, education, and the environment under "Policy First" and "Sustainability First" "without sacrificing economic development in most regions." Inclusion of input from all levels of society in the "Sustainability First" scenario leads to greater buy-in; it reduces degradation, strengthens local rights, and builds capacity and legitimacy.

This example makes it clear that our existing policies and institutional arrangements are inadequate to deal with current environmental problems.[3] Innovative, flexible, and diversified policy options are required. We need to value the jobs, wealth, and social benefits that our market system creates. We need to recognize where the things we consume come from and how they are produced, and we must value the environment from which they are obtained. We need to recognize that poverty breeds increased environmental degradation, economic instability, and politic unrest. At the same time, we must not let our enthusiasm for change cause us to lose sight of the wonderful contributions our current system has provided and can continue to provide.

Our governments need to show leadership. They need to act on their intergenerational responsibilities. Their policies must recognize the economic, health, and security impact of environmental change because the economy, health, security, and environment are intrinsically linked. Businesses are looking to government for leadership and regulation to provide long-term stability for capital investments and changes in business practices. The sooner government provides these, the sooner businesses will contribute their wealth, knowledge, and innovation to making positive economically and environmentally sustainable changes. Governments need to facilitate progressive industries and their initiatives. Governments must act because failure to do so will affect and define generations to come. Determined action must begin immediately, as the cost of waiting far exceeds that of immediate action. As individuals, communities, governments, policy makers, and business leaders, we must seek new ideas from diverse peoples and organizations and recognize that much greater success will result if we work together toward global sustainability instead of competing against one another.

By recognizing our symbiotic relationship with each other and other species on the planet, we will be better able to reflect on our own identity and mutual dependency. Such a perspective will engender greater care in our behavior toward each other. We will accept a new perspective in which diversity is celebrated and survival becomes neither the sole privilege nor the consequence of dominance.

It is also clear that a technological "fix" will not be sufficient to "cure" the problem. As Neanderthals found, a technological change must be matched with a coincident behavioral shift. The automobile industry provides a good modern-day example of the importance of behavioral change. When the U.S. government, concerned about the country's dependence on foreign oil and rising fossil fuel prices, put in place a new Energy Policy and Conservation Act in 1975, the auto industry improved motor efficiency and reduced vehicle air resistance. However, consumers offset these benefits by purchasing new fuel-consuming features for their vehicles, including air conditioners and other electronic options, and drove with fewer passengers in their cars.[4] Gas consumption went up.

If consumers do not "buy in" and shift their behavior in response to government and business initiatives, nothing will change. Thus, a shift from fossil fuel technology to some other form of fuel is not likely to mitigate future climate change. Excessive consumption will simply generate a new problem elsewhere. For example, in the case of biofuels, excessive use will intensify rising demand for available cropland, which is also needed to grow food for an expanding population in a world where rising temperatures and limited freshwater resources will reduce the viability of agricultural land. It will also lead to further deforestation, exacerbating rising atmospheric CO_2 levels.

Government initiatives combined with technological advances are absolutely necessary, but they are not sufficient. Not all innovation and change will necessarily be good, nor will change for change's sake. For example, technological fixes based on conflict over cooperation risk leading us into a siege mentality that will result in the adoption of "aggressive surveillance, tracking, communication and warfare technologies. We risk choosing surveillance and militarization over human rights. We risk choosing cell-phone tracking systems and remote-control and robotic weapons for borders."[5] Technological advances that control or dominate nature, other species, or humans must be developed and implemented with extraordinary care because it is through domination and control that we have generated the economic and environmental mess in which we currently find ourselves. Conflict brings with it extreme risks – the escalation of nuclear and biological warfare and the possible demise of *Homo sapiens* and many other species. Those who choose cooperation over confrontation will have a greater chance of adjusting and avoiding war, famine, and poverty. Cooperation means embracing diversity; it likely also means we will have to consume less and share more.

What is desperately needed is a multipronged initiative that emerges on all fronts, from governments and business to schools, religious institutions, and societies, from the community to the family, right down to the individual. It is individuals who will display the most immediate and profound leadership in change. We as individuals need to take responsibility because every entity, whether it is political, economic, religious, educational, social, or cultural, consists of individuals. Each and every one of us, in every region of the world, must discover how, as individuals and groups of individuals, we are contributing to the problem and in what ways we can contribute to its solution. There will be as many alternative solutions as there are varied contributors.

We must begin by dwelling in unity – recognizing that we are all in this together – and continue by experiencing diversity – recognizing that each of us have unique contributions to make. We are united in that what we do influences others and what they do influences us; as a species we are supported by the complexity of life on this planet. Yet we are diverse – there will be many solutions, but we must remain open and tolerant in order to recognize the varied ways of dealing with this complex problem. It will take many different approaches and perspectives to develop a viable, effective solution.

As we begin to implement our various strategies, I believe that we will not only solve our climate change problem, but we will also contribute to the understanding and resolution of other global issues including child poverty, HIV and AIDS, violence against women, and other social disorders. We sit on the precipice of change. A world divided denotes disaster, but a world united is a harbinger of hope. We are set to make the greatest contributions *Homo sapiens* has yet made for Earth – it is your choice.

I continue to belt out "City of New Orleans," and the CBC radio network recently rated "This Land Is Our Land" a classic. Despite natural selection, Huntington's discordant mutations continue unabated. Woody and his family may be considered unfit by neo-Darwinists, but their songs and convictions provide hope for future generations.

Grandmothers are raising millions of children orphaned by AIDS and HIV in Africa. Ordinary people, many of them women, alike in that many have been operating on the fringes of the dominant culture, are finding hope and making positive contributions that far exceed their initial expectations. They survive despite being different – perhaps because they are different. They are engaging in positive action; they will leave Earth a better place for having lived here for a fraction of geological time. For many it is the single most important thing they are doing in their lives. They are making a difference, and so can you.

Dare to let go of the fear, discover your hope, and find the strength and courage to act on it. Listen to these words of a wonderful mentor, Dr. Ursula Franklin, and live in a dangerous climate.

> If you look at the world as it is now, nobody in their right mind intended it to be like that and it is not solely lack of foresight, of which there is a lot, but also genuine lack of understanding of how things work. And that is not stupidity, but it is the limitation of what the human mind can grasp.... It wasn't that someone single handedly delivered the mess in a plain brown envelope. It evolved as nature's response to step-by-step stupid decisions and responses not read. It is as if nature constantly sent emails and nobody opened them of those that could.
>
> Humans are only a very small part of nature. In certain ways humans function best the better they understand the workings of nature and the more they respect it. They have the right, as every creature has, to modify nature so that they survive. That is why some birds migrate and some birds stay.... And that is a perfectly rightful cycle in nature. Now it doesn't modify the climate, it modifies one's behaviour in order to live in the existing climate, and that is a very great difference, understanding nature so that one can modify one's own behaviour and survive.[6]

Glossary

Abu Hureyra — first founded 12,700 years ago, this is one of the most famous archaeological sites in the Fertile Crescent, where agriculture first developed between 11,000 and 8,000 years ago. Abundant wild cereals, domesticated rye, emmer, einkorn wheat, and barley have been found at this site.

Acheulean — a stone toolkit made by *Homo erectus* and *Homo heidelbergensis* from approximately 1.5 million to 200,000 years ago, first found at the site of St. Acheul in France but since uncovered throughout Europe, Africa, and east into the Indian subcontinent; most noted for its large, tear-drop-shaped, two-sided handaxes and cleavers, which disappeared in Europe 500,000 years ago but not until 200,000 years ago elsewhere.

Adaptability — the ability of an organism to effectively adjust itself to a changing internal or external environment.

Adaptation — a genetically fixed and inflexible trait that is appropriate to particular internal or external environments.

Agriculturalist — someone who cultivates the soil to grow crops.

Albedo — the amount of solar radiation that comes into the Earth's atmosphere from the sun that is immediately reflected back, either

	by clouds, atmospheric gases, aerosols, or the Earth's surface. A higher planetary albedo occurs when more of the Earth's land surface is covered in snow relative to trees.
Amino acid	the building blocks of proteins; hundreds of different amino acids exist, but all proteins that are known in organisms consist of long chains of the same twenty amino acids.
Anthropologist	someone who studies humanity, namely human origins, the biological and cultural development of humans, past and present.
Archaeologist	someone who studies humanity's cultural past by studying past materials and artifacts including tools, artwork, and debris found in archaeological sites.
Atlantic meridional overturning (AMO)	a conveyor-like circulation of Atlantic Ocean waters that brings warm surface water northward. Upon reaching northern latitudes, those warm waters then cool and are driven down to depth forming cold, deep waters that then flow southward. AMO transports heat from low latitudes to high latitudes in the North Atlantic.
Aurignacian	Upper Palaeolithic toolkit with beads, bone carvings, and stone tools believed to have been made by early *Homo sapiens* in western, central, and eastern Europe beginning around 40,000 years ago and remaining after Châtelperronian tools had stopped being produced; also associated with cave art such as in Chauvet cave in southern France.
Australopithecus species	meaning "southern ape," an early, now extinct, early genus of bipedal hominins that emerged approximately 4 million years ago or earlier and believed to be related to humans; includes the species listed below (see also Table 2.1).
Australopithecus	lived 3.85 million to 2.95 million years ago

afarensis	in eastern Africa. Its best-known representative is the skeleton named "Lucy," who was 1 meter tall, weighed about 27 kilograms, and had a brain less than one-third the size of that in modern humans.
Australopithecus africanus	lived 3.3 million to 2.1 million years ago in southern Africa. With their ape- and human-like features they were very similar to *A. afarensis*, although slightly taller and heavier.
Australopithecus anamensis	"anam" is "lake" in Turkana language; bipedal hominin believed to be transitional between apes and later australopithecines; dates to between 4.2 million and 3.9 million years ago.
Australopithecus robustus	also called *Paranthropus robustus*, a robust *Australopithecus* that had a sagittal crest – a ridge of bone that ran along the top of its head from front to back; first found at Kromdraai cave, Africa, and lived between 1.8 million to 1.2 million years ago.
Australopithecus aethiopicus	another robust *Australopithecus* found at Lake Turkana, Africa, and lived 2.5 million years ago; famous for the "black skull" that experienced mineral uptake from manganese-rich soil staining it blue-black.
Australopithecus sediba	sediba means "fountain" in Sesotho language of South Africa, most recent find of *Australopithecus sp.* made by Lee Berger in South Africa of two partial skeletons that are between 1.98 million and 1.78 million years old, respectively; had long arms for climbing but also walked upright.
Beringia	the continental shelf that extends from northeastern Siberia to the Mackenzie River in Yukon, Canada, which was exposed during the last ice age when global average sea levels dropped 120 meters below today's levels, forming the "Beringian landbridge" and along which early people are thought to have lived and migrated.

Bifacial tool — a two-sided stone tool produced by chipping or flaking a stone cobble or flake on both sides; examples first appeared in Oldowan tool kit, also Acheulean and Clovis stone tools; includes, among others, tools such as bifacial points (arrow and spear heads), bifacial hand axes, bifacial scrapers for butchering, scraping, and cleaning animal hides.

Bipedal — walking on two legs or rear limbs.

Bolide — a large meteor that explodes when it passes through Earth's atmosphere.

Châtelperronian — Upper Palaeolithic stone tools from Europe, believed to have been made by early *Homo sapiens*, although the finding of 36,000-year-old Neanderthal skeletal remains at Saint-Césaire, France, alongside Châtelperronian stone tools has cast doubt on this interpretation.

Chromosomes — rod-like bodies in the nucleus made up of DNA; humans have twenty-three pairs, including two sex chromosomes, for a total of forty-six; during reproduction, chromosomes are replicated with each parent providing twenty-three chromosomes to the offspring.

Climate — (as opposed to weather - weather is not climate, weather refers to the atmospheric conditions in the short term, what you see outside your window today) climate is the probability of that weather continuing to occur - over the long term.

Clovis — a biface stone tool technology first discovered near Clovis and Folsom, New Mexico, in the 1920s and 1930s, dated to as far back as 13,400 years ago.

Clovis people — the name given to the people who made the Clovis stone tool technology; believed to be big-game hunters who followed the massive ice-age mammals across the Beringian landbridge; initially believed to be the first people to inhabit North America; also known as Paleoindians.

Codon a unit of genetic coding; consists of three nucleotides in a DNA or RNA molecule, which form into a single amino acid.

Cognitive requirement in reference to the level of cognitive, intellectual, and creative ability deemed necessary to perform and fulfill a knowledge-intensive task like creating symbols and art.

Cranium (*pl.* crania) the skull of a vertebrate; encloses the brain.

Cro-Magnon considered to be an early *Homo sapiens* that lived about 35,000 years ago in Europe, known for beautiful cave paintings in France and Spain, made tools to hunt and kill prey, lived in caves and rock shelters; had a cranial capacity of 1,400 cc, averaged 1.8 meters tall, had a high forehead, prominent chin, and no brow ridges.

Darwinian theory of evolution Charles Darwin's theory of evolution of species through the action of natural selection and competition for survival by the fittest by which genetically determined characteristics change gradually over generations.

Deforestation to clear of forest or trees.

Dentition teeth, their number, arrangement, and type in a given species.

Dispersal event to separate and go in different directions, to move into new territories – for example, the *Homo* dispersal event out of Africa 900,000 years ago.

DNA deoxyribonucleic acid – the principal hereditary genetic material of an organism, situated in the cell's nucleus and organized in chromosomes; two long strands of nucleotides form the double helix of DNA; it consists of thousands of genes which are the instructions for building proteins that perform cell functions; DNA replicates itself (with the help of enzymes).

Domestication	the taming of wild animals to live with humans; in reference to agriculture, the cultivation of wild plants.
Dominant species	the predominant, or prevailing, species in a plant or animal community; retains priority access to food and other resources.
Dwarf *Stegodon*	a primitive dwarf elephant first believed to have gone extinct 840,000 years ago; however, the recent find associated with *Homo floresiensis* on Flores Island suggests a new dwarf species developed that survived until 17,000 years ago.
Dynamic stasis	in reference to evolution, a time when evolutionary changes are rare or resisted by prevailing conditions; species' form and distribution is relatively stable and resistant to change.
Ecosystem	a community of biological organisms that interact with each other and their physical environment.
El Niño–Southern Oscillation (ENSO) events	a somewhat periodic and prolonged warming in the tropical Pacific Ocean waters and change in currents first described in 1892. Originally, El Niño was thought to be a local weather pattern bringing warm temperatures and rain to the coast of Peru as opposed to the normal cool and dry climate. Now it is regarded as a manifestation of a global phenomenon called the Southern Oscillation. El Niño remains the common name, although it is formally known as El Niño–Southern Oscillation (ENSO). The alternate cooling phase is known as La Niña.
Emergence theory of evolution	the sudden appearance of something new, either through saltatory emergence (e.g., the symbiotic evolution that occurred when mitochondria entered host cells) or critical-point emergence (e.g., the evolution of flight), may be directly caused by environmental change, although is

	ultimately gene-based; spurts of emergent evolution are followed by long intervals of stasis where little change occurs and natural selection presides. (For more on this subject, see R. G. B. Reid [2007]. *Biological Emergences: Evolution by Natural Experiment.* Cambridge, MA: The MIT Press, pp.78–94).
Epidemiologist	someone who studies the incidence and distribution of diseases and how to control and prevent them.
Equilibrium	a state of balance between physical forces; in the case of climate modeling, a state where the Earth system is in balance - in other words, the amount of incoming solar energy equals the amount of energy emitted into space. For example, an increase in atmospheric CO_2 concentration limits the energy emitted into space and will generate a response and coincidental change in atmospheric temperature. The climate model helps determine what that new equilibrium climate will be like and how long it will take for the Earth's system to reach equilibrium after such a change.
Eskimo-Aleut	maritime hunters and speakers of the Eskimo (Inuit and Yupik) and Aleut languages who are believed to have initially expanded out of eastern Siberia and occupied the Aleutian chain of islands and the lower half of the Alaska Peninsula and later expanded across much of Alaska, the Canadian Arctic, Nunavik, Nunatsiavut, and Greenland. (For more on the early archaeological history of these people, see West, F. H., ed. [1996]. *American Beginnings: The Prehistory and Palaeoecology of Beringia.* Chicago: University of Chicago Press.)
Eukaryotic cells	a type of cell with a true nucleus.
Evolution	a process in which the form and overall genetic structure of an organism changes over generations.

Fertile Crescent	a region of fertile land in the Middle East that extends from the east Mediterranean coast across the Tigris and Euphrates river valleys to the Persian coast; a region also known as the "cradle of civilization" because it was the home of early Sumerian, Phoenician, Babylonian, and Assyrian civilizations.
Fossil	anything ancient, especially if it is buried, such as bones, stone tools, tracks, habitation sites, and so on.
Fossil fuels	fuels created by geological processes that concentrated the energy of earlier organisms after they died and which can be burned for fuel, such as oil, coal, and gas.
Genome	the structure within the organism that contains the instructions for the organism's growth and reproduction; the collection of genes that determine an organism's makeup – the DNA.
Genomist	a person who studies the genetic information of organisms.
Germ line	consists of the cells that produce the sexual gametes, the sperm and eggs used in sexual reproduction to produce a next generation.
Gigatons	one billion tons.
Great apes	hominids; the original hominid line diverged about 6 million years ago into two branches; one branch split into chimpanzees and gorillas, the second emerged approximately 4 million years ago or earlier into *Australopithecus sp.*
Greenhouse gases	certain gases in Earth's atmosphere, including methane, carbon dioxide, nitrous oxide, and water, that act like glass in a greenhouse, allowing visible light to reach the surface of Earth but trapping infrared energy from escaping into the outer atmosphere, and thereby trapping heat.

Haploid/diploid	having only a single set of chromosomes, one copy of the genome, like a sperm or egg cell; as compared with diploid, having a double set of chromosomes, two copies of the genome; for example, the zygote has two copies of its genome, one from each parent.
Heat-shock protein	helps maintain basic cell functions including cell temperature and chemical balance; when an organism is under stress, heat-shock proteins also can change methylation patterns, thereby stimulating the expression of normally repressed genes.
Hominid	refers to all the great apes both modern and extinct, see *Great apes*.
Hominin	refers to the bipedal apes including all the fossil species and living humans.
Homo antecessor	some scientists consider *H. antecessor* the potential common ancestor of both Neanderthals and ourselves; appeared about 800,000 to 900,000 years ago in Spain and possibly Italy.
Homo erectus	"upright human," potential *Homo sapiens*' direct ancestor that appeared in Africa about 1.89 million years ago and lived until about 143,000 years ago and perhaps as late as 50,000 years ago in Java – about nine times longer than *Homo sapiens* have been on Earth; had a brain size between 700 and 1,250 cc, a body size similar to humans, made and used tools and fire.
Homo ergaster	some scientists consider this a separate species from *Homo erectus* that originated in Lake Turkana, Kenya, 1.9 million years ago and remained in Africa to become the direct ancestor of *Homo sapiens*; had a brain size up to 850 cc.
Homo floresiensis	the last known *Homo* species to have lived, except for humans; a 1-meter-tall relative with a brain size of 380 cc, lived on Flores Island, Indonesia from between about

	95,000 to 17,000 years ago when it went extinct; long-term isolation on Flores Island is considered an explanation for its small size, although based on 1 million-year-old tools, others suggest it evolved from *Homo erectus*, or possibly another small early species of the genus *Homo*, and came to the island a million years ago or more.
Homo habilis	"handy" or "able human," thought to be the first known hominin to use complex stone tools called the Oldowan technology; lived 2.4 million to 1.4 million years ago on an omnivorous diet in grasslands by lakes and rivers in Africa; brain size 600 to 800 cc.
Homo heidelbergensis	lived 700,000 to 200,000 years ago in Europe, Africa and possibly China; had a larger brain size than humans, controlled and used fire, and built dwellings out of rock and wood.
Homo neanderthalensis	Neanderthals, separated from the human lineage about 500,000 years ago; migrated to Europe and Asia long before *Homo sapiens*; the last known Neanderthal lived in Portugal about 24,500 years ago; their skull was longer and lower than *Homo sapiens* with noticeably heavy brow and receded chin; they lived in caves and shelters, made weapons and tools, and cared for their dead; their brain size was about 1,450 cc.
Homo sapiens	first appeared about 200,000 years ago; believed to have evolved in Africa before moving out of Africa 120,000 years ago. Compared with *Homo erectus*, had a slightly larger brain size (1,300–1,400 cc – about the same as today) and rounder skull, smaller teeth and jaw, high forehead, prominent chin, and no brow ridges.
Hopeful monster	a concept of geneticist Richard Goldschmidt referring to offspring that are born radically different from their parents; hopeful in that they may find an appropriate mate

	with which they could breed and generate fertile offspring; based on his idea that major structural changes can occur without a series of small intermediate changes.
Huntington's disease	a genetic disease caused by a defect in chromosome 4; passed down through families; affects muscles and cognitive ability, worsens over time; self-amplifying nature of the codon repetition means that the disease will affect subsequent generations earlier.
Hypothesis	an idea or concept that provides a basis for explanation or arguments that can be tested by experimentation.
Ice core	a sample of ice taken by drilling into it; typically obtained from ice sheets in Antarctica and Greenland; used by scientists to understand paleoenvironments.
Indonesian Orang Pendek	"Little Man of the Forest"; a half-human half-ape creature that is part of the folklore of the Indonesian tribespeople.
Industrial Revolution	the dramatic shift in society that occurred when much of the working population left agriculture and began to work in industries, particularly in Britain, during the late 1700s and early to mid-1800s.
Inferior left frontal lobe	located at the left front of the brain (cortex); the frontal lobes are responsible for voluntary motor activity, speaking, and elaboration of thought.
Interglacial	a period of warmer climate between two glacial periods or ice ages, based on evidence that shows a characteristic shift in vegetation from colder to warmer species.
Interstadial	a period of warmer climate within a glacial period that is shorter, and thought to be less warm, than an interglacial.
Intertropical convergence zone – ITCZ	also known as the doldrums where the northeasterly trade winds from the north meet the southeasterly tradewinds from the south and cause showers, thunderstorms,

	and hurricanes; it is responsible for the wet and dry seasons in the tropics; it typically migrates northward during the northern hemisphere's summer and southward reaching just south of the equator during the winter.
Isotope	a chemical element that has two or more varieties insofar as they have a common atomic number, but the varieties have a different number of neutrons in their nucleus; for example, the hydrogen in water and the hydrogen in "heavy water" are two different hydrogen isotopes.
Kennewick Man	a 9,500-year-old skeleton found along the Columbia River, Washington, with skull characteristics similar to Caucasian people and dental characteristics similar to Southeast Asians and Jomon-Ainu people of Japan.
Komodo dragon	an ancient lizard species that first appeared 40 million years ago; today, as the world's largest lizard, it can reach up to 3 m (10 feet) in length and weigh up to 126 kg (277 lbs); found in Indonesia, shared the island of Flores with *Homo floresiensis* from 95,000 to 17,000 years ago.
Levallois	a stone-toolmaking technology that involves a series of steps whereby flakes are struck from a prepared core to create stone tools; particularly evident during the Middle Palaeolithic.
Levant	a region that roughly stretches across modern Israel, Lebanon, Syria, Jordan, and the West Bank.
Marsupial	a mammal of the Order Marsupialia; born partially developed, it completes its development in a pouch on the mother's belly where it suckles. Example: kangaroo.
Megafauna	animals weighing more than 45 kilograms, many of which prevailed during the last ice age and subsequently became extinct.

Mesoamerica	the central region of the Americas, from Mexico to Nicaragua.
Methylation/ methylation patterns	the process by which certain genes are turned on or off, allowing cells and organs to specialize and perform specific functions. Normal methylation patterns can be altered by extreme stress caused by high temperatures, exposure to oxidants, infection, and abnormal metabolic rates.
Middle Palaeolithic tool industry	the stone-tool industry of early *Homo* species dating from about 250,000 to 40,000 years ago; for example, Mousterian.
Middle Stone Age tool industry	prevalent in Africa around 285,000 years ago and somewhat later in Europe; new smaller, more diverse tools appeared; a key innovation included the "prepared core technology," such as Levallois stone-toolmaking technology; tools included well-produced handaxes, small points used for projectile weapons, and stone awls and scrapers for puncturing and preparing hides.
Milankovic cycles	three distinct cycles in Earth's orbit around the Sun that profoundly affect variations in solar radiation; first calculated by Serbian mathematician and planetary physicist Milutin Milankovic; see Chapter 7 for a description.
Mitochondria	organelles in eukaryotic cells that produce adenosine triphosphate (ATP), the chemical energy source of a cell, from foodstuffs.
Mitochondrial DNA (mtDNA)	the DNA in the energy-making mitochondria found within human cells; contains a small fraction of our total DNA and is passed from a mother to her offspring
Morphological transition	a change in the shape, color, or other general feature of an organism.
Mousterian stone tool industry	a Middle Palaeolithic tool industry made by Neanderthals in which flake tools were detached from prepared cores.
Movius line	a geographic line that separates the region where, to the west, hand axes were

	manufactured, from a region to the east, where they were not (see Figure 5.3).
Mutation	a small but stable change in the structure of the DNA molecules that form the genome, which alters the information that the genome carries, or alters the expression of that information.
Na-Dené	speakers of the Na Dené language; believed to have originated from a forest-adapted people from Siberia, thought to have expanded solely into North America.
Natural selection	a theory that individuals having the greatest *fitness*, those with the genetic type within a species that leave the greatest number of offspring, are more likely to pass those genes on and successfully compete and survive in nature.
Neocortex	the outer layer of grey matter that forms part of the brain of mammals; in larger mammals, it is grooved and folded, providing greater surface area without increasing the volume of the brain; in humans, it is enlarged and is responsible for advanced cognitive skills including speech and language.
Nonrepetitive fine motor skills	small muscle movements, particularly of the fingers, usually in coordination with the eyes, that are performed without repetition – for example, writing.
Nucleotides	made up of a sugar, nitrogenous base and a phosphate, which, when condensed into chains, form nucleic acids – DNA and RNA.
Oasis theory	anthropologist V. G. Childe's theory that a colder, dryer climate forced humans and animals into "oases" where there remained good, reliable sources of water and food, and that the concentration of growing populations in these centers led to the development of agriculture.
Occipital lobes	the lobes of the brain that are located at the back of the head; responsible for initially processing what the eyes see.

Oldowan technology	a stone toolkit made and used by *Homo habilis*; the oldest known stone tool technology; first discovered in the Olduvai Gorge in Tanzania; contains stone "choppers" used for cutting plants and scraping and butchering animals.
Omnivorous diet	a varied diet that includes many kinds of food, including meat, fish, and plants.
Opposable thumb	primates possess this thumb, which operates on a different plane than do the other digits and can be used to press against the other digits to allow grasping.
Overkill	to kill or destroy beyond that which is necessary, or appropriate, for food or victory.
Oxidants	a substance that causes the oxidation of another substance; a reaction in which oxygen combines with, or hydrogen is removed from, a substance, causing an atom to lose an electron.
Ozone	in reference to the ozone layer, a layer of concentrated ozone that blankets the Earth at an altitude of 15 to 30 km; it limits the amount of ultraviolet radiation that reaches Earth's surface.
Palaeontologist	someone who specializes in the study of extinct and fossil animals and plants.
Paleoanthropologist	someone who specializes in the study of fossil human remains.
Paleo-Arctic people	a forest-adapted people who are believed to have migrated from Siberia to North America and who were the ancestors of the Na-Dené.
Paleoindian	see Clovis people.
Palaeolithic	the "old" stone age divided into the Lower, Middle, and Upper Palaeolithic; includes stone tools and other artifacts that began with the earliest stone toolmaking at least 2.6 million years ago and culminated about 10,000 years ago.
Pan paniscus	the bonobo chimpanzee, found in central Africa, is more slightly built than chimpanzees, shows a greater preference for walking

	on two legs than other apes, and eats mainly fruit, seeds, and plants (but does consume mammals, although does not actively hunt them); bonobos are peaceful compared with the more violence-prone chimpanzee.
Paranthropus	an early, now extinct, bipedal hominin – for example, *Paranthropus robustus*; also called *Australopithecus robustus*.
Pastoralist	someone who raises and tends livestock.
Periodicity	the time taken to complete one cycle; the time taken for an event to recur.
Photosynthesis	the process by which energy derived from sunlight is used by organisms, particularly green plants, to create carbohydrates from CO_2 and water; it generates oxygen as a by-product.
Phylum (*pl.* Phyla)	groups of organisms; for example, the phylum Chordata includes animals with a notochord at the back of the brain.
Physical anthropologist	an anthropologist who studies humans as a biological species.
Placental animals	animals whose young develop inside the mother's uterus attached to a placenta, which provides for absorption of food, dissipation of waste, and a supply of oxygen; the young of placental animals develop to a greater maturity before birth than do the young of other mammals.
Precision grip	a delicate grip of the hand used by chimpanzees and humans to manipulate small objects.
Radial core technology	a stone tool technology in which flakes are removed from a core and used to make tools.
RNA – ribonucleic acid	a single-stranded nucleic acid that contains the sugar ribose; genes are transcribed into RNA, processed using messenger RNA (mRNA), transported out of the cell using transfer RNA (tRNA), and used to build proteins.

Saltatory process	leaps in evolution as opposed to slow gradual evolution with its many intermediate forms.
Savanna	a grassy plain with few or no trees; found in tropical and subtropical regions.
Sequential nonrepetitive fine motor control	the coordination of small muscles, bones, and nerves, particularly of the fingers, to perform precise movements without repetition and sequentially.
Sexual gamete - or germ cell	one of the reproductive cells of a sexually producing organism, for example the egg or sperm that fuse during fertilization to produce a zygote that develops into another adult; contains one set of chromosomes.
Sink (carbon sink)	the ability of the biosphere and oceans to drawdown CO_2 out of the atmosphere.
Social evolution	the ability of a society to change its behavior and characteristics.
Species	a group of closely related organisms; in sexual organisms, populations of organisms that are capable of interbreeding with each other and producing fertile offspring; in asexually producing organisms, those that have a common ancestry and whose characteristics are relatively stable and of a similar type.
Speciation	the process by which a new species is formed.
Stadials	a period of increased cold or advancing ice within a glacial period.
Steppic regions	a plain that is relatively flat grassy and unforested.
Subcontinent	used in reference to Beringia to describe the exposed continental shelf that stretched between northeast Asia and northwest North America during the last ice age.
Theory	a supposition, idea, or group of systematic ideas explaining something, for example the theory of evolution.

Thermohaline circulation	the movement of large bodies of water such that cool surface waters move vertically downward, forcing warm waters upward (overturning); this causes mixing of the water; in oceans this usually involves both changes in temperature and salinity.
Trade winds	prevailing tropical winds that blow from the subtropics to the northern hemisphere in a typically northeasterly direction and to the southern hemisphere in a generally southwesterly direction.
Transitional form	a form that is a passing state from one form to another; a stage of progression.
Unilinear progression	in which species evolve along a single path without branching into multiple species.
Upwelling	in large bodies of water, the wind-generated movement of cold, nutrient-rich subsurface waters from depth to the surface.
Upper Palaeolithic tool industry	the stone tool industry of *Homo sapiens* dating from about 40,000 years ago to 10,000 years ago; a diverse, higher-quality, and more refined technology than the Middle Palaeolithic, with new tools like the atl-atl for throwing projectiles and beautiful "venus" figurines and cave paintings.
Y-chromosomes	sex chromosomes in male animals; males have one x-chromosome and one y-chromosome, females have two x-chromosomes.
Zygote	a cell formed by the union of one egg and one sperm during sexual reproduction; each egg and sperm provide half the number of chromosomes for the zygote, which has a full complement of chromosomes; it develops into an adult.

Notes

1. PUTTING OUR EMERGENT HOUSE IN ORDER

1. Brinkhuis, H., Schouten, S., Collinson, M. E., Sluijs, A., Damsté, J. S. S., et al., and the Expedition 302 Scientists (2006). Episodic fresh surface waters in the Eocene Arctic Ocean. *Nature*, 441, 606–9.
2. Such impacts appear in the geological record as a major biological change. The steady accumulation of sediments over a long period of climatic and biological stability create a uniform rock strata, which, after sudden climatic change, is abruptly disrupted as old species disappear from the fossil record and new species appear. For an example of research on volcanism, see Rampino, M. R., Self, S., and Stothers, R. B. (1988). Volcanic winters. *Annual Review of Earth and Planetary Science*, 16, 73–99, and Rampino, M. R., and Stothers, R. B. (1988). Flood basalt volcanism during the past 250 million years. *Science*, 241, 663–8.
3. See, for example, Keller, G., Adatte, T., Pardo Juez, A., and Lopez-Oliva, L. (2009). New evidence concerning the age and biotic effects of the Chicxulub impact in Mexico. *Journal of the Geological Society*, 166, 393–411.
4. Switek, B. (2011). "Jurassic Mother" found in China. *Science*NOW (August 24), http://www.sciencemag.org/
5. For a complete discussion of the concept of stasis, see Reid, R. G. B. (2007). *Biological Emergences: Evolution by Natural Experiment*, Vienna Series in Theoretical Biology. Cambridge, MA: MIT Press, chapter 8, especially pp. 326–8.
6. The term *hominin* here refers to the bipedal apes, including all the fossil species and living humans, whereas *hominid* refers to all the great apes.
7. For a good example of the effects of changing sea level on the capacity of early humans to inhabit and migrate along Canada's west coast during and after the last ice age, see Hetherington, R., Barrie, J. V., MacLeod, R., and Wilson, M. (2004). Quest for the lost land. *Geotimes*, 49, 20–3.
8. For a discussion and explanation of how, when, and why, see Chapter 6 and Hetherington, R. and Reid, R. G. B. (2010). *The Climate Connection: Climate Change and Modern Human Evolution*. Cambridge: Cambridge University Press.
9. Lüthi, D., Le Floch, M., Bereiter, B., Blunier, T., Barnola, J-M., et al. (2008). High-resolution carbon dioxide concentration record 650,000–800,000 years before present. *Nature*, 453, 379–82; Petit, J. R., Jouzel, J., Raynaud, D., Barkov, N. I., Barnola,

J.-M., et al. (1999). Climate and atmospheric history of the past 420,000 years from the Vostok ice core, Antarctica. *Nature*, 399, 429–36.

10. For a nice discussion on the impacts of production and technology, see Franklin, U. M. (2006). The real world of mathematics, science, and technology education. In *The Ursula Franklin Reader: Pacifism as a Map*. Toronto: Between the Lines, pp. 328–35.
11. Although it is thought that aboriginal peoples had permanently inhabited the Canadian Arctic by 6,000 years ago – and lived there intermittently earlier – and that hunter-gatherers inhabited desert regions in Africa many thousands of years ago, our pervasive expansion into these environments over the past 150 years has been made possible by fossil fuel technology.
12. See Petit, J. R. et al. (1999). Climate and atmospheric history of the past 420,000 years from the Vostok Ice Core, Antarctica. *Nature*, 399, 429–36, figure 2.
13. Marland, G., Boden, T. A., and Andres, R. J. (2005). Global, regional, and national CO_2 emissions. In *Trends: A Compendium of Data on Global Change*. Oak Ridge, TN: Carbon Dioxide Information Analysis Center, Oak Ridge National Laboratory, U.S. Department of Energy.
14. Siegenthaler, U., Stocker, T. F., Monnin, E., Lüthi, D., Schwander, J., et al. (2005). Stable carbon cycle-climate relationship during the Late Pleistocene. *Science*, 310, 1313–7.
15. Chen, C., Hill, J. K., Ohlemüller, R., Roy, D. B., and Thomas, C. D. (2011). Rapid range shifts of species associated with high levels of climate warming. *Science*, 333, 1024–6.
16. For a discussion of these ideas and the three C's, see Hetherington, R., and Reid, R. G. B. (2010). *The Climate Connection: Climate Change and Modern Human Evolution*. Cambridge: Cambridge University Press.

2. THE CRADLE OF HUMANKIND

1. Quoted in Butler, S. (1879). *Evolution Old and New*. London: Hardwicke and Bogue, 90–1.
2. Darwin, C. (1871). *The Descent of Man, and Selection in Relation to Sex*, 2 vols. London: Murray.
3. Huxley, T. H. (1863). *Evidence as to Man's Place in Nature*. London: Williams and Norgate.
4. Potts, R. (1996). *Humanity's Descent: The Consequences of Ecological Instability*. New York: Morrow, 8.
5. For a nice summary of the history of this and other early hominins, see the Smithsonian National Museum of Natural History's Web site "What Does It Mean to Be Human?" at http://humanorigins.si.edu/
6. Dennell, R. W., Rendell, H., and Hailwood, E. (1988). Early tool-making in Asia: two-million year-old artefacts in Pakistan. *Antiquity*, 62, 98–106; Rendell, H. M., Hailwood, E. A., and Dennell, R. W. (1987). Magnetic polarity stratigraphy of Upper Siwalik Sub-Group, Soan Valley, Pakistan: Implications for early human occupance of Asia. *Earth and Planetary Science Letters*, 85, 488–96.
7. Gabunia, L., Vekua, A., Lordkipanidze, D., Swisher, C. C., Ferring, R., Justus, A., Nioradze, M., Tvalchrelidze, M., Antón, S. C., Bosinski, G., Jöris, O., de Lumley, M-A., Majsuradze, G., and Mouskhelishvili, A. (2000). Earliest Pleistocene hominid cranial remains from Dmanisi, Republic of Georgia: Taxonomy, geological

setting, and age. *Science*, 288, 1019–25; Gabunia, L., Antón, S. C., Lordkipanidze, D., Vekua, A., Justus, A., and Swisher, C. C. (2001). Dmanisi and dispersal. *Evolutionary Anthropology*, 10, 158–70.

8. Larick, R., Ciochon, R. L., Zaim, Y., Sudijono, Suminto, Rizal, Y., Aziz, F., Reagan, M., and Heizler, M. (2001). Early Pleistocene 40Ar/39Ar ages for Bapang Formation hominins, Central Jawa, Indonesia. *Proceedings of the National Academy of Sciences of the United States of America*, 98, 4866–71.
9. Ambrose, S. H. (2001). Paleolithic technology and human evolution. *Science*, 291, 1748–53; Asfaw, B., Beyene, Y., Suwa, G., Walter, R. C., White, T. D., WoldeGabriel, G., and Yermane, T. (1992). The earliest Acheulean from Konso-Gardula. *Nature*, 360, 732–5; Clark, J. D. (1994). *The Acheulian industrial complex in Africa and elsewhere.* In *Integrative Paths to the Present*, eds. R. S. Corruccini and R. L. Ciochon. Englewood Cliffs, NJ: Prentice-Hall, 451–69.
10. Gibbons, A. (2001). Tools show humans reached Asia early. *Science*, 293, 2368–9; Zhu, R. X., Hoffman, K. A., Potts, R., Deng, C. L., Pan, Y. X., Guo, B., Shi, C. D., Guo, Z. T., Yuan, B. Y., Hou, Y. M., and Huang, W. W. (2001). Earliest presence of humans in northeast Asia. *Nature*, 413, 413–17.
11. Hou, Y., Potts, R., Baoyin, Y., Zhengtang, G., Deino, A., et al. (2000). Mid-Pleistocene Acheulean-like stone technology of the Bose basin, South China. *Science*, 287, 1622–6.
12. Wood, B., and Collard, M. (1999). The human genus. *Science*, 284, 65–71.
13. Asfaw, B., Gilbert, W. H., Beyene, Y., Hart, W. K., Renne, P. R., et al. (2002). Remains of *Homo erectus* from Bouri, Middle Awash, Ethiopia. *Nature*, 416, 317–20.
14. Specifically, they are controlled by areas of the brain's inferior left frontal lobe; Greenfield, P. M. (1991). Language, tools and brain: The ontogeny and phylogeny of hierarchically organized sequential behavior. *Behavioral and Brain Sciences*, 14, 531–95; Kempler, D. (1993). Disorders of language and tool use: Neurological and cognitive links. In *Tools, Language and Cognition in Human Evolution*, eds. K. R. Gibson and T. Ingold. Cambridge: Cambridge University Press, 193–215.
15. Wood, B., and Collard, M. (1999). The human genus. *Science*, 284, 65–71.
16. Pääbo, S. (2003). The mosaic that is our genome. *Nature*, 421, 409–12.
17. Britten, R. J. (2002). Divergence between samples of chimpanzee and human DNA sequences is 5%, counting indels. *Proceedings of the National Academy of Sciences of the United States of America*, 99, 13633–5.
18. Chakravarti, A. (2001). Single nucleotide polymorphisms … to a future of genetic medicine. *Nature*, 409, 822–3; Sachidanandam, R., Weissman, D., Schmidt, S. C., Kakol, J. M., Stein, L. D., et al., and International SNP Map Working Group (2001). A map of human genome sequence variation containing 1.42 million single nucleotide polymorphisms. *Nature*, 409, 928–33.
19. The neocortex is the outer layer of grey matter that forms part of the brain of mammals. In larger mammals it is grooved and folded, providing greater surface area without increasing the volume of the brain. In humans, the neocortex is enlarged and is responsible for advanced cognitive skills including speech and language.
20. Tattersall, I. (2003). Once we were not alone. *Scientific American*, 13, 20–7.
21. This is the hypothesis put forward by Hetherington, R., and Reid, R. G. B. (2010). *The Climate Connection: Climate Change and Modern Human Evolution*. Cambridge: Cambridge University Press.

3. THE NEANDERTHAL ENIGMA

1. Mellars, P. (2005). The impossible coincidence: a single-species model for the origins of modern human behavior in Europe. *Evolutionary Anthropology*, 14, 12–27.
2. Ibid.
3. Mercier, N., Valladas, H, Joron, J. L., Reyss, J. L., Lévêque, F., and Vandermeersch, B. (1991). Thermoluminescence dating of the late Neanderthal remains from Saint-Césaire. *Nature*, 351, 737–9.
4. Hetherington, R., and Reid, R. G. B. (2010). *The Climate Connection: Climate Change and Modern Human Evolution*. Cambridge: Cambridge University Press.
5. For a slightly different perspective, see Banks, W. E., d'Errico, F., Peterson, A. T., Kageyama, M., Sima, A., and Sánchez-Goñi, M-F. (2008). Neanderthal extinction by competitive exclusion. *PLoS ONE* 3(12), e3972. doi10.1371/journal.pone.0003972. The authors suggest that the habitat niche *H. sapiens* occupied in Europe during Marine Isotope Stage 3 was different than that of Neanderthals. They suggest that although both habitats initially shrank during MIS3 (60,000 to 30,000 years ago), after about 40,000 years ago, the habitat in which *H. sapiens* lived in Europe expanded, whereas the Neanderthals' habitat shrank considerably. As a result, *H. sapiens* expanded into Neanderthals' territory, successfully competed with them, and caused Neanderthals to go extinct.
6. For a good discussion on this subject, see Mellars, P. (1996). The big transition. In *The Neanderthal Legacy: An Archaeological Perspective from Western Europe*. Princeton, NJ: Princeton University Press, pp. 392–419.
7. Green, R. E., Drause, J., et al. (2010). A draft sequence of the Neanderthal genome. *Science*, 328, 710–22. For a nice description of the scientific findings, see Callaway, E. (2010). Neanderthal genome reveals interbreeding with humans. *New Scientist*, May 6, 2010, http://www.newscientist.com/article/dn18869-neanderthal-genome-reveals-interbreeding-with-humans.html
8. Finlayson, C. (2004). *Neanderthals and Modern Humans: An Ecological and Evolutionary Perspective*. Cambridge: Cambridge University Press.
9. Martin, P. S. (1973). The discovery of America. *Science*, 179, 969–74; Ellis, C., Goodyear, A. C., Morse, D. F., and Tankersley, K. B. (1998). Archaeology of the Pleistocene-Holocene transition in eastern North America. *Quaternary International*, 49/50, 151–66.
10. See Flannery, T. (1995). *Mammals of New Guinea*. Sydney: Australian Museum/Reed Books; Miller, G. H., Magee, J. W., Johnson, B. J., Fogel, M. L., Spooner, N. A., et al. (1999). Pleistocene extinction of *Genyornis newtoni*: Human impact on Australian megafauna. *Science*, 283, 205–8; Roberts, R. G., Flannery, T. F., Ayliffe, L. K., Yoshida, H., Olley, J. M., et al. (2001). New ages for the last Australian megafauna: Continent-wide extinction about 46,000 years ago. *Science*, 292, 1888–92.

4. THE END OF *HOMO* DIVERSITY

1. Morwood, M. J., O'Sullivan, P. B., Aziz, F., and Raza, A. (1998). Fission-track ages of stone tools and fossils on the east Indonesian island of Flores. *Nature*, 392, 173–6; Brumm, A., Jensen G. M., et al. (2010). Hominins on Flores, Indonesia, by one million years ago. *Nature*, 464, 748–52.

2. For a nice description of this, see McKie, R. (2010). How a hobbit is rewriting the history of the human race. *Guardian*, February 21, http://www.guardian.co.uk/science/2010/feb/21/hobbit-rewriting-history-human-race
3. Paul Broca's "Sur le volume et la forme du cerveau suivant les individus et suivant les races" (1861), quoted in Gould, S. J. (1981). *The Mismeasure of Man.* New York: W.W. Norton, 83.
4. Not all scientists agree that *Homo floresiensis* was a separate species. Rather, some think it was a dwarfed version of *Homo sapiens.* If this were the case, the last *Homo* relative to have lived would be *Homo neanderthalensis*, which disappeared perhaps as late as 24,500 years ago.
5. However, N. Rolland and S. Crockford suggest that prior to this recent discovery, it was thought that the dwarf *Stegodon*, represented by *Stegodon sondaarii*, went extinct 840,000 years ago, before *Homo floresiensis* arrived on the island. It may be that another dwarf *Stegodon* developed subsequently or this was a juvenile of a normal sized *Stegodon*. Rolland, N., and Crockford, S. (2005). Late Pleistocene dwarf Stegodon from Flores, Indonesia? *Antiquity*, 79, http://antiquity.ac.uk/projgall/rolland/index.html
6. The Orang Pendek ("Little Man of the Forest") is a half-human, half-ape creature that is part of the folklore of the Indonesian tribespeople.

5. CLIMATE AND HUMAN MIGRATION

1. For a nice visual summary of the changing vegetation in Africa, see "Africa during the last 150,000 years," compiled by Jonathan Adams, Environmental Sciences Division, Oak Ridge National Laboratory, http://www.esd.ornl.gov/projects/qen/nercAFRICA.html
2. Dennell, R. W. (2004). Hominid dispersals and Asian biogeography during the Lower and Early Middle Pleistocene, c. 2.0–0.5 Mya. *Asian Perspectives*, 43, 205–26; Hetherington, and Reid (2010). *The Climate Connection.*
3. On vegetation, see Faure, H., Walter, R. C., and Grant, D. R. (2002). The coastal oasis: Ice age springs on emerged continental shelves. *Global and Planetary Change*, 33, 47–56; Hetherington, R., Wiebe, E., Weaver, A. J., Carto, S., Eby, M., and MacLeod, R. (2008). Climate, African and Beringian subaerial continental shelves, and migration of early peoples. *Quaternary International*, 183, 83–101. On shellfish, see Crawford, M. A., Bloom, M., Broadhurst, C. L., Schmidt, W. F., Cunnane, S. C., Galli, C., Gehbremeskel, K., Linseisen, F., Lloyd-Smith, J., and Parkington, J. (1999). Evidence for the unique function of docosahexaenoic acid during the evolution of the modern hominid brain. *Lipids*, 34, 39–47; Foley, R. A. (2002). Adaptive radiations and dispersals in hominin evolutionary ecology. *Evolutionary Anthropology*, Supplement 1, 32–7; Klein, R. G., Avery, G., Cruz-Aribe, K., Halkett, D., Hart, T., Milo, R. G., and Volman, T. P. (1999). Duinefontein 2: An Acheulean site in the Western Cape Province of South Africa. *Journal of Human Evolution*, 37, 153–90.
4. Faure, Walter, and Grant (2002). The coastal oasis; Hetherington, Wiebe, Weaver, Carto, Eby, and MacLeod (2008). Climate, African and Beringian subaerial continental shelves, and migration of early peoples.
5. Bar-Yosef, O., and Belfer-Cohen, A. (2001). From Africa to Eurasia – early dispersals. *Quaternary International*, 75, 19–28; MacPhee, R. D., and Marx, P. A. (1997). The 40,000-year plague: Humans, hyperdisease, and first-contact extinctions. In

Natural Change and Human Impact in Madagascar, eds. S. M. Goodman and B. D. Patterson. Washington, DC: Smithsonian Institution Press, pp. 169–217.

6. For a discussion of these ideas, see Hetherington, R., and Reid, R. G. B. (2010). *The Climate Connection: Climate Change and Modern Human Evolution*. Cambridge: Cambridge University Press.
7. McBrearty, S., and Brooks, A. S. (2000). The revolution that wasn't: A new interpretation of the origin of modern human behavior. *Journal of Human Evolution*, 39, 453–563.
8. Petit, J. R., Jouzel, J., Raynaud, D., Barkov, N. I., Barnola, J.-M., Basile, I., Bender, M., Chappellaz, J., Davis, M., Delayque, G., Delmotte, M., Kotlyakov, V. M., Legrand, M., Lipenkov, V. Y., Lorius, C., Pépin, L, Ritz, C., Saltzman, E., and Stievenard, M. (1999). Climate and atmospheric history of the past 420,000 years from the Vostok ice core, Antarctica. *Nature*, 399, 429–36.
9. Gathorne-Hardy, F. J., and Harcourt-Smith, W. E. H. (2003). The super-eruption of Toba, did it cause a human bottleneck? *Journal of Human Evolution*, 45, 227–30.
10. Van den Bergh, G. D., Meijer, H. J. M., Rokhus Due Awe, Morwood, M. J., Szabó, K., van den Hoek Ostende, L. W., Sutikna, T., Saptomo, E. W., Piper, P. J., and Dobney, K. M. (2009). The Liang Bua faunal remains: A 95 k.yr. sequence from Flores, East Indonesia. *Journal of Human Evolution*, 57, 527–37.
11. Forster, P. (2004). Ice ages and the mitochondrial DNA chronology of human dispersals: A review. *Philosophical Transactions of the Royal Society of London, Series B, Biological Sciences*, 359, 255–64; Oppenheimer, S. (2003). Journey of mankind interactive trail adapted from Out of Eden/The Real Eve, www.bradshawfoundation.com/journey/
12. Macaulay, V., Hill, C., Achilli, A., Rengo, C., Clarke, D., Meehan, W., Blackburn, J., Semino, O., Scozzari, R., Cruciani, F., Taha, A., Shaari, N. K., Raja, J. M., Ismail, P., Zainuddin, Z., Goodwin, W., Bulbeck, D., Bandelt, H-J., Oppenheimer, S., Torroni, A., and Richards, M. (2005). Single, rapid coastal settlement of Asia revealed by analysis of complete mitochondrial genomes. *Nature*, 308, 1034–6.
13. Hetherington, R., and Reid, R. G. B. (2010). *The Climate Connection*.
14. Roberts et al. (1990) and Roberts et al. (1994) have suggested that behaviorally modern human occupation of Australia occurred by about 55,000 years ago. Bowler, Johnston, et al. (2003) came up with new optical ages that date the human occupation at Lake Mungo III, 2,700 kilometres from the current northwestern Australian coast, which was originally dated by Thorne et al. (1999) to 62,000 ± 6,000 years ago. These new dates are between 50,000 and 45,000 years ago for the initial occupation and 40,000 ± 2,000 years ago for two human burials. A review of recent research by O'Connell and Allen (2004), resulted in more conservative dating for the colonization of Pleistocene Australia and New Guinea at between 45,000 and 42,000 years ago. Roberts, R. G., Jones, R., and Smith, M. A. (1990). Thermoluminescence dating of a 50,000-year-old human occupation site in northern Australia. *Nature*, 345, 153–6; Roberts, R. G., Jones, R., and Smith, M. A. (1994). Beyond the radiocarbon barrier in Australia. *Antiquity*, 68, 611–16; Bowler, J. M., Johnston, H., Olley, J. M., Prescott, J. R., Roberts, R. G., Shawcross, W., and Spooner, N. A. (2003). New ages for human occupation and climatic change at Lake Mungo, Australia. *Nature*, 421, 837–40; Thorne, A., Grün, R., Mortimer, G., Spooner, N. A., Simpson, J. J., McCulloch, M., Taylor, L., and

Curnoe, D. (1999). Australia's oldest human remains: Age of the Lake Mungo 3 skeleton. *Journal of Human Evolution*, 36, 591–612; O'Connell, J. F., and Allen, J. (2004). Dating the colonization of Sahul (Pleistocene Australia–New Guinea): A review of recent research. *Journal of Archaeological Science*, 31, 835–53.

15. Miller, G. H., Magee, J. W., Johnson, B. J., Fogel, M. L., Spooner, N. A., McCulloch, M. T., and Ayliffe, L. K. (1999). Pleistocene extinction of *Genyornis newtoni*: Human impact on Australian megafauna. *Science*, 283, 205–8; Hetherington and Reid (2010), *The Climate Connection*, p. 224.
16. Hopkins, M. (2005). Early African migrants made eastward exit. *Nature*, 12, May 2005, doi: 10.1038/news050509-10; Hetherington, R., and Reid, R. G. B. (2010). *The Climate Connection*.
17. For an excellent discussion of this earlier African transition, see McBrearty, S., and Brooks, A. S. (2000). The revolution that wasn't: A new interpretation of the origin of modern human behaviour. *Journal of Human Evolution*, 39, 453–563.

6. BRAVING THE NEW WORLD

1. For a description of the impact of glaciations on Canada's west coast and what it meant for early human migrants to the Americas, see Hetherington, R., Barrie, J. V., MacLeod, R., and Wilson, M. (2004). Quest for the lost land. *Geotimes*, 49, 20–3.
2. Guthrie, R. D. (2001). Origin and causes of the mammoth steppe: A story of cloud cover, woolly mammal tooth pits, buckles, and inside-out Beringia. *Quaternary Science Reviews*, 20, 135–47; Kaplan, J. O. (2001). Geophysical Applications of Vegetation Modeling [PhD thesis]. Lund, Sweden, Lund University, 114 pp; Dyke, A. S. (2004). An outline of North America deglaciation with emphasis on central and northern Canada. In *Quaternary Glaciations-Extent 40 and Chronology, Part II North America*, eds. J. Ehlers and P. L. Gibbard. Amsterdam: Elsevier, Developments in Quaternary Science, 2b, 373–424; Hetherington, R., Wiebe, E. Weaver, A. J., Carto, S., Eby, M., and MacLeod, R. (2008). Climate, African and Bergingian subaerial continental shelves, and migration of early peoples. *Quaternary International*, 183, 83–101.
3. This is an excerpt from the Tlingit story Kák'w Shaadaax' x'éidáx sh kalneek (Basket Bay History), as told by Shaadaax' (Robert Zuboff) and contributed by Naatstláa (Constance Naish) and Shaachooká (Gillian Story) as a memorial to Shaadaax'. It tells how the Tlingit migrated along the Pacific Northwest coast of North America and how their population expanded. Today the Tlingit live in southeast Alaska, southwest Yukon, and northwest British Columbia. The story is transcribed in Dauenhauer, N. M., and Dauenhauer, R., eds. (1987). *Classics of Tlingit Oral Literature, Vol. 1: Haa Shuká, Our Ancestors Tlingit Oral Narratives*. Seattle: University of Washington Press, pp. 67–71.
4. See Heusser, C. J. (1960). *Late Pleistocene Environments of Pacific North America*, Special Publication 35. New York: American Geological Society; Krieger, A. D. (1961). Review of "Late Pleistocene environments of North Pacific North America," by C. J. Heusser (1960). *American Antiquity*, 27, 249–50; Macgowan, K., and Hester, J. A. Jr. (1962). *Early Man in the New World*. New York: Doubleday.
5. Fladmark, K. R. (1979). Routes: Alternate migration corridors for early man in North America. *American Antiquity*, 44, 55–69.

6. See Rogers, R. A. (1985). Glacial geography and native North American languages. *Quaternary Research*, 23, 130–7; Rogers, R. A. (1985). Wisconsinan glaciation and the dispersal of native ethnic groups in North America. In *Woman, Poet, Scientist: Essays in New World Anthropology Honoring Dr. Emma Lou Davis*, ed. T. C. Blackburn. Los Altos, CA: Ballena Press, pp. 104–13; also see Rogers, R. A., Martin, L. D., and Nicklas, T. D. (1990). Ice Age geography and the distribution of native North American languages. *Journal of Biogeography*, 17, 131–43.
7. See Gruhn, R. (1988). Linguistic evidence in support of the coastal route of earliest entry into the New World. Man (N.S.), 23, 77–100; Gruhn, R. (1994). The Pacific coast route of entry: An overview. In *Method and Theory for Investigating the Peopling of the Americas*, eds. R. Bonnichsen and D. G. Steele. Corvallis: Center for the Study of the First Americans, Oregon State University, pp. 249–56.
8. For a discussion of the human bone, see Dixon, E. J. (2001). Human colonization of the Americas: Timing, technology and process. *Quaternary Science Reviews*, 20, 277–99; for a thorough description of the early fauna at this site, see Heaton, T. H., Talbot, S. L., and Shields, G. F. (1996). An ice age refugium for large mammals in the Alexander Archipelago, southeastern Alaska. *Quaternary Research*, 46, 186–92.
9. For an in-depth discussion of this research, including paleogeographic maps of the area since the last ice age, see Hetherington, R., and Reid, R. G. B. (2003). Malacological insights into the marine ecology and changing climate of the late Pleistocene–early Holocene Queen Charlotte Islands archipelago, western Canada, and implications for early peoples. *Canadian Journal of Zoology*, 81, 626–61; Hetherington, R., Barrie, J. V., Reid, R. G. B., MacLeod, R., and Smith, D. J. (2004). Paleogeography, glacially induced crustal displacement, and Late Quaternary coastlines on the continental shelf of British Columbia, Canada. *Quaternary Science Reviews*, 23, 295–318; Hetherington, R., Barrie, J. V., MacLeod, R., and Wilson, M. (2004). Quest for the lost land. *Geotimes*, 49, 20–3.
10. These bones were originally found by Phil C. Orr in the early 1960s at Arlington Springs, Santa Rosa Island. See Orr, P. C. (1962). Arlington Springs Man. *Science*, 19, 219. Cited in Dixon, E. J. (1999). *Bones, Boats and Bison: Archeology and the First Colonization of Western North America*. Albuquerque: University of New Mexico Press, p. 129.
11. First published by Keefer, D. K., deFrance, S. D., Moseley, M. E., Richardson, III, J. B., Satterlee, D. R., and Day-Lewis, A. (1998). Early maritime economy and El Niño events at Quebrada Tacahuay, Peru. *Science*, 281, 1833–5.
12. See Stanford, D., and Bradley, B. (2000). The Solutrean solution - did some ancient Americans come from Europe? *Scientific American Discovering Archaeology*, 2, 54–5; Stanford, D., and Bradley, B. (2002). Ocean trails and prairie paths? In *The First Americans: The Pleistocene Colonization of the New World*, ed. N. G. Jablonski. Berkeley: University of California Press, pp. 255–71; also see Holden, C. (1999). Archaeology: Were Spaniards among the first Americans? *Science*, 286, 1467–8. For an opposing view, see Straus, L. G. (2000). Solutrean settlement of North America? A review of reality. *American Antiquity*, 65, 219–26.
13. Montenegro, A., Hetherington, R., Eby, M., and Weaver, A. J. (2006). Modelling pre-historic transoceanic crossings into the Americas. *Quaternary Science Reviews*, 25, 1323–38.
14. See Neves, W. A., Powell, J. F., and Ozolins, E. G. (1999). Extra-continental morphological affinities of Lapa Vermelha IV, Hominid 1: A multivariate analysis

with progressive numbers of variables. *Homo*, 50, 263–82; Neves, W. A., Prous, A., González-José, R., Kipnis, R., and Powell, J. (2003). Early Holocene human skeletal remains from Santana do Riacho, Brazil: Implications for the settlement of the New World. *Journal of Human Evolution*, 45, 19–42; see also Lahr, M. M. (1995). Patterns of modern human diversification: Implications for Amerindian origins. *Yearbook of Physical Anthropology*, 38, 163–98.

15. See Neves, W. A., Powell, J. F., and Ozolins, E. G. (1999). Modern human origins as seen from the peripheries. *Journal of Human Evolution*, 34, 96–105; Steele, D. G., and Powell, J. F. (2002). Facing the past: A view of the North American human fossil record. In *The First Americans: The Pleistocene Colonization of the New World*, ed. N. G. Jablonski. Berkeley: University of California Press, pp. 93–122. For a contrary view, see Van Vark, G. N., Kuizenga, D., and L'Engle Williams, F. (2003). Kennewick and Luzia: Lessons from the European Upper Paleolithic. *American Journal of Physical Anthropology*, 121, 181–4; but for a response, see Neves, W. A., Prous, A., González-José, R., Kipnis, R., and Powell, J. (2003). Early Holocene human skeletal remains from Santana do Riacho, Brazil: Implications for the settlement of the New World. *Journal of Human Evolution*, 45, 19–42.
16. See Hurtado de Mendoza, D., and Braginski, R. (1999). Y Chromosomes point to Native American Adam. *Science*, 283, 1439–40.
17. See Kemp, B. M., Malhi, R. S., McDonough, J., Bolnick, D. A., Eshleman, J. A., et al. (2007). Genetic analysis of early Holocene skeletal remains from Alaska and its implications for the settlement of the Americas. *American Journal of Physical Anthropology*, 132, 605–21; for a nice discussion, see also Dalton, R. (2005). Caveman DNA hints at map of migration. *Nature*, 436, 162.
18. Montenegro, A., Hetherington, R., Eby, M., and Weaver, A. J. (2006). Modelling pre-historic transoceanic crossings into the Americas. *Quaternary Science Reviews*, 25, 1323–38.
19. Betty Meggers has produced a lifetime of work on Jomon and other early pottery. For an example, see Meggers, B. J. (1971). The transpacific origin of Mesoamerican civilization: A preliminary review of the evidence and its theoretical implications. *American Anthropology*, 77, 1–27; Meggers, B. J. (1998a). Archaeological evidence for transpacific voyages from Asia since 6000 BP. *Estudios Atacameños*, 15, 107–24; Meggers, B. J. (1998b). Jomon-Valdivia similarities: Convergence or contact? In *Across before Columbus? Evidence for Transoceanic Contact with the Americas Prior to 1492*, eds. D. Y. Gilmore and L. S. McElroy. Edgecomb, ME: New England Antiquities Research Association (NEARA) Publications, pp. 16–17; Meggers, B. J. (1998c). Archaeological evidence for transpacific voyages from Asia since 6000 BP. *Estudios Atacameños*, 15, pp. 107–20. Nancy Davis has written a wonderful book that delves into many of the cultural links described here: Davis, N. Y. (2001). *The Zuni Enigma*. New York: Norton. For an article on the Chinese stone anchors, see Pierson, L. J., and Moriarty, J. R. (1980). Stone anchors: Asiatic shipwrecks off the California coast. *Anthropological Journal of Canada*, 18, 17–23.

7. AGRICULTURE AND THE RISE OF CIVILIZATION

1. Milankovic, M. (1998). *Canon of Insolation and the Ice-Age Problem*. Royal Serbian Academy, special publications, v. 132.
2. First described in 1892, El Niño was originally thought to be a local weather pattern bringing warm temperatures and rain to the coast of Peru, as opposed to the

normal cool and dry climate. Now it is regarded as a manifestation of a global phenomenon called the Southern Oscillation. El Niño remains the common term, although it is formally known as El Niño–Southern Oscillation (ENSO).

3. For an up-to-date measurement of global average sea level and related references, see the chart of the "Global Mean Sea Level Time Series" on the CU Sea Level Research Group Web site (http://sealevel.colorado.edu/).
4. Belyaev, D. K. (1979). Destabilizing selection as a factor in domestication. *Journal of Heredity*, 70, 301–8.
5. Lief Andersson is the research group leader of this work. See Giuffra, E., Kijas, J. M. H., Amarger, V., Carlborg O., Jeon J.-T., Andersson, L. (2000). The origin of the domestic pig: Independent domestication and subsequent introgression. *Genetics*, 154, 1785–91.
6. Fagan, B. (2004). *The Long Summer: How Climate Changed Civilization.* New York: Basic Books.
7. Turney, C. S. M., and Brown, H. (2007). Catastrophic early Holocene sea level rise, human migration and the Neolithic transition in Europe. *Quaternary Science Reviews*, 26, 2036–41.
8. For a complete discussion of the domestication of peanuts, cotton, and squash, see Dillehay, T. D., Rossen, J., Andres, T. C., and Williams, D. E., (2007). Preceramic adoption of peanut, squash, and cotton in Northern Peru. *Science*, 316, 1890–3.
9. For the original report on this, see Pope, K. O., Pohl, M. E. D., Jones, J. G., Lentz, D. L., von Nagy, C., Vega, F. J., and Quitmyer, I. R. (2001). Origin and environmental setting of ancient agriculture in the lowlands of Mesoamerica. *Science*, 292, 1370–3.
10. See Barber, D. C., Dyke, A., Hillaire-Marcel, C., Jennings, A. E., Andrews, J. T., Kerwin, M. W., Bilodeau, G., McNeely, R., Southon, J., Morehead, M. D., and Gagnon, J.-M. (1999). Forcing of the cold event of 8,200 years ago by catastrophic drainage of Laurentide lakes. *Nature*, 400, 344–8; Turney, C. S. M., and Brown, H. (2007). Catastrophic early Holocene sea level rise, human migration and the Neolithic transition in Europe. *Quaternary Science Reviews*, 26, 2036–41.
11. These findings are based on work published in 2010 by R. Hetherington and R. G. B. Reid, drawn from their analysis of 135,000 years of climate, vegetation, and archaeological evidence, in their book *The Climate Connection: Climate Change and Modern Human Evolution* (Cambridge University Press).
12. For a more in-depth analysis of the origin of agriculture in South Asia, see Allchin, B., and Allchin, R. (1997). *Origins of a Civilization: The Prehistoric and Early Archaeology of South Asia.* New Delhi: Penguin Books; and Gupta, A. K. (2004). Origin of agriculture and domestication of plants and animals linked to early Holocene climate amelioration. *Current Science*, 87, 54–9.
13. These findings are based on work published in 2010 by R. Hetherington and R. G. B. Reid on their analysis of 135,000 years of climate, vegetation, and archaeological evidence in their book: *The Climate Connection: Climate Change and Modern Human Evolution* (Cambridge University Press).
14. For a thorough discussion of the onset of agriculture and civilization, see Diamond, J. (1999). *Guns, Germs, and Steel: The Fates of Human Societies.* New York: Norton; and Fagan, B. (2004). *The Long Summer.*
15. Ryan, W., and Pitman, W. (1999). *Noah's Flood. The New Scientific Discoveries about the Event that Changed History.* New York: Simon and Schuster. This publication first made this story public, and although it generated heated controversy, the

subsequent scientific account by Turney and Brown is very persuasive. Turney, C. S. M., and Brown, H. (2007). Catastrophic early Holocene sea level rise, human migration and the Neolithic transition in Europe. *Quaternary Science Reviews*, 26, 2036–41.

16. Hole, F. (1994). *Environmental Instabilities and Urban Origins: Chiefdoms and Early States in the Near East. The Organizational Dynamics of Complexity*, Monographs in World Archaeology 18. Madison, WI: Prehistory Press.
17. See Brooks, N. (2006). Cultural responses to aridity in the Middle Holocene and increased social complexity. *Quaternary International*, 151, 29–49.
18. See Fagan, B. (2004). *The Long Summer*, for a discussion.
19. Hetherington, R., and Reid, R. G. B. (2010). *The Climate Connection*.
20. Di Lernia, S. (2006). Building monuments, creating identity: Cattle cult as a social response to rapid environmental changes in the Holocene Sahara. *Quaternary International*, 151, 50–62.
21. For the destruction of neighboring habitat, see Dow, G. K., Olewiler, N., and Reed, C. G. (2005). The transition to agriculture: Climate reversals, population density, and technical change. Simon Fraser University, Economic Discussion Paper. For more on the diffusion of agriculture, see Diamond, J. (1997). *Guns, Germs, and Steel: The Fates of Human Societies*. New York: Norton.
22. See Fagan, B. (2004). *The Long Summer*; also Brooks, N. (2006). Cultural responses to aridity..., p. 32.
23. Jousse, H. (2006). What is the impact of Holocene climatic changes on human societies? Analysis of West African Neolithic populations dietary customs. *Quaternary International*, 151, p. 71.

8. THE MAYA CIVILIZATION AND BEYOND

1. Hodell, D. A., Curtis, J. H., and Brenner, M. (1995). Possible role of climate in the collapse of classic Mayan civilisation. *Nature*, 375, 341–7; cited in Fagan, B. (2004). *The Long Summer: How Climate Changed Civilization*. New York: Basic Books, p. 236.
2. Haug, G. H., Günther, D., Peterson, L. C., Sigman, D. M., Hughen, K. A., and Aeschlimann, B. (2003). Climate and the collapse of Maya civilization. *Science*, 299, 1731–5.
3. Fagan, B. (2004). *The Long Summer*, p. 237.
4. First described in 1892, El Niño was first thought to be a local weather pattern bringing warm temperatures and rain to the coast of Peru as opposed to the normal cool and dry climate. Now it is regarded as a manifestation of a global phenomenon called the Southern Oscillation. El Niño remains the common name, although it is formally known as El Niño–Southern Oscillation (ENSO).
5. Baker, P. A., Seltzer, G. O., Fritz, S. C., Dunbar, R. B., Grove, M. J., Tapia, P. M., Cross, S. L., Rowe, H. D., and Broda, J. P. (2001). The history of South American tropical precipitation for the past 25,000 years. *Science*, 291, 640–3.
6. Brooks, N. (2006). Cultural responses to aridity in the Middle Holocene and increased social complexity. *Quaternary International*, 151, 29–49.
7. Burroughs, W. J. (2005). *Climate Change in Prehistory: The End of the Reign of Chaos*. Cambridge: Cambridge University Press.
8. See Brooks, N. (2006). Cultural responses to aridity... pp. 44–5, for a nice discussion on this topic.

9. DOMINANCE DESTABILIZED

1. As described in Chapter 1 of this book, dynamic stasis is a relatively unchanging state in which organisms, although resistant to change, experience "minor natural experiments [leading] to minor adaptations, without achieving any [evolutionary] progress." For a complete discussion of the concept of "stasis," see Reid, R. G. B. (2007). *Biological Emergences: Evolution by Natural Experiment*, Vienna Series in Theoretical Biology. Cambridge, MA: MIT Press, chapter 8, especially pp. 326–8.
2. Hsiang, S. M., Meng, K. C., and Cane, M. A. (2011). Civil conflicts are associated with the global climate. *Nature*, 476, 438–41.

10. FITNESS FOLLY

1. Darwin, C. (1959). *On the Origin of Species*. New York/Boston: H. M. Caldwell Company, p. 38. Originally published in 1859.
2. Elizabeth Dodson Gray, quoted in Maggio, R. (1992). *The Bias-Free Word Finder.* Boston: Beacon, p. 195.
3. C. P. Estés (1992) does a wonderful job of illuminating the concept of the "too-good mother" in chapter 3, "Nosing out the facts: The retrieval of intuition as initiation," in *Women Who Run with the Wolves: Myths and Stories of the Wild Woman Archetype*. New York: Ballantyne Books, pp. 74–114.

11. DARWIN THE SELECTOR

1. Darwin, C. (1959). *On the Origin of Species*. New York/Boston: H. M. Caldwell Company, pp. 447, 450. Originally published in 1859.
2. See Eldredge, N., and Gould, S. J. (1972). Punctuated equilibria: An alternative to phyletic gradualism. In *Models in Paleobiology*, ed. T. J. M. Schopf. San Francisco: W. H. Freeman, Cooper; also Gould, S. J., and Eldredge, N. (1993). Punctuated equilibrium comes of age. *Nature*, 366, 223–7; Eldredge, N. (1995). *Reinventing Darwin: The Great Debate at the High Table of Evolutionary Theory.* New York: Wiley and Sons; and Gould, S. J. (1997). *The Mismeasure of Man*. Penguin: Harmondsworth.

12. HUNTING DOWN WOODY

1. Fondon, J. W., III, and Garner, H. R. (2004). Molecular origins of rapid and continuous morphological evolution, *Proceedings of the National Academy of Sciences*, 101, 18058–63.
2. Reid, Robert G. B. (2007). *Biological Emergences: Evolution by Natural Experiment.* Cambridge, MA: MIT Press, p. 53.
3. The idea that people choose their partners based on genes may seem unbelievable to some readers, particularly those whose partners are of the same sex or who married for reasons other than having children. We like to think that our choices are based on something beyond the molecular level, but people do choose spouses for the physical expression of their genes, whether they are blond, blue-eyed, and statuesque (characteristics that are genetically determined), have a sense of humor or a good singing voice, are dynamite on the dance floor, boast a big wallet, and so on. However, the simplistic view that people pick each other for their good genes highlights the dangers associated with a reductionist perspective. When we break everything down to the smallest genetic component, we can lose

sight of the holistic and interdependent nature of organisms and our capacity to choose and modify our behavior.

13. KAMMERER'S SUICIDE

1. For an example, see Chapter 15, "Dutch Hunger Winter Babies."

14. GIANTS AND PYGMIES

1. Darwin, C. (1959). *On the Origin of Species.* New York/Boston: H. M. Caldwell Company, p. 38. Originally published in 1859.
2. Ferretti, G., Atchou, G., Grassi, B., Marconi, C., Cerretelli, P. (1991). Energetics of locomotion in African pygmies. *European Journal of Applied Physiology*, 62, 7–10.
3. Garcia-Cao, I. et al. (2002). "Super p53" mice exhibit enhanced DNA damage response, are tumor resistant and age normally. *EMBO Journal*, 21, 6225–35; Greenwood, E. (2002). Research highlight: Cancer prevention: Supermice. *Nature Reviews Cancer*, 2, 889 (doi:10.1038nrc957). See also The supermice that resist cancer. BBC News, November 2, 2004, www.news.bbc.co.uk/2/hi/health/3971103.stm

15. DUTCH HUNGER WINTER BABIES

1. Lumey, L. H. (1992). Decreased birthweights in infants after maternal in utero exposure to the Dutch famine of 1944–1945. *Paediatric and Perinatal Epidemiology*, 6, 240–53; Lumey, L. H. (1998). Reproductive outcomes in women prenatally exposed to undernutrition: A review of findings from the Dutch famine birth cohort. *Proceedings of the Nutrition Society*, 5, 129–35; note also that it is the female parent who passes on the expressional response to a changing environment. This highlights the importance of the female as the harbinger of change.
2. Sollars, V., Lu, X., Xiao, L., Wang, X., Garfinkel, M. D., and Ruden, D. M. (2002). Evidence for an epigenetic mechanism by which Hsp90 acts as a capacitor for morphological evolution. *Nature Genetics*, 33, 70–4; see also Silva, A. J., and White, R. (1988). Inheritance of allelic blueprints for methylation patterns. *Cell*, 54, 145–52.
3. For a nice discussion on this work done by Randy Jirtle and Robert Waterland at Duke University, see Watters, E. (2006). DNA is not destiny. *Discover: Science, Technology, and The Future*, November, http://discovermagazine.com/2006/nov/cover
4. Jablonka, E. (2004). Epigenetic epidemiology. *International Journal of Epidemiology*, 33, 929–35. See also Wikipedia, s.v. "Dutch Famine of 1944," en.wikipedia.org/wiki/Dutch_famine_of_1944; and Jeffrey, S. (1995) The high price of hunger. *Medical Post*, November 7, www.mentalhealth.com/mag1/p5m-sc01.html. For a good general biological reference for heat-shock proteins, see Pirkkala, L., Nykanen, P., and Sistonen, L. (2001). Roles of the heat shock transcription factor in regulation of the heat shock response and beyond. *FASEB Journal*, 15, 1118–31.

16. TODAY AND TOMORROW

1. Much of the scientific data presented in this chapter has been obtained from three main sources: Intergovernmental Panel on Climate Change (IPCC) (2007). *Climate Change 2007: The Physical Science Basis.* Contribution of Working Group I to

the Fourth Assessment Report of the IPCC, ed. S. Solomon, D. Qin, M. Manning, Z. Chen, M. Marquis, K. B. Averyt, M. Tignor, and H. L. Miller. Cambridge: Cambridge University Press, available online at ipcc-wg1.ucar.edu/wg1/wg1-report.html; IPCC (2007). *Climate Change 2007: Impacts, Adaptation and Vulnerability.* Contribution of Working Group II to the Fourth Assessment Report of the IPCC, ed. M. L. Parry, O. F. Canziani, J. P. Palutikof, P. J. van der Linden, and C. E. Hanson. Cambridge: Cambridge University Press, available online at www.gtp89.dial.pipex.com/chpt.htm; Warren, R. (2006). Impacts of global climate change at different annual mean global temperature increases. In *Avoiding Dangerous Climate Change*, ed. H. J. Schellnhuber, W. Cramer, N. Nakicenovic, T. Wigley, and G. Yohe. Cambridge: Cambridge University Press, pp. 93–131.

2. For an up-to-date source of current CO_2 levels, see the CO_2Now.org Web site, http://co2now.org/
3. For a continuously updated global mean sea level measurement using satellite radar and tide gauges, along with links to other sources, see the CU Sea Level Research Group Web site, http://sealevel.colorado.edu/; see also Brahic, C. (2007). Sea level rise outpacing key predictions. *New Scientist Environment*, February 1, www.environment.newscientist.com/article/dn11083
4. Goswami, B. N., Venugopal, V., Madhusoodanan, M. S., Xavier, P. K. (2006). Increasing trend of extreme rain events over India in a warming environment. *Science*, 314, 1442–5.
5. Weaver, A. J., Saenko, O. A., Clark, P. U. and Mitrovica, J. X. (2003). Meltwater pulse 1A from Antarctica as a trigger of the Bølling-Allerød Warm Interval. *Science*, 299, 1709–13.
6. Eby, M., Zickfield, K., Montenegro, A., Archer, D., Meissner, K. and Weaver, A. (2009). Lifetime of anthropogenic climate change: Millennial time scales of potential CO_2 and surface temperature perturbations. *Journal of Climate*, 22, 2501–11; http://geosci.uchicago.edu/~archer/reprints/eby.2009.long_tail.pdf
7. Arnell, N., and Liu, C. (2001). "Hydrology and water resources," chapter 4 in *Climate Change 2001: Impacts, Adaptation and Vulnerability,* Contribution of Working Group II to the Third Assessment Report of the IPCC, ed. J. J. McCarthy, O. F. Canziani, N. A. Leary, D. J. Dokken, and K. S. White. Cambridge: Cambridge University Press, p. 213, http://www.grida.no/publications/other/ipcc_tar/?src=/climate/ipcc_tar/wg2/180.htm
8. Although the most recent IPCC 2007 report estimates sea level will rise at least 7 meters over the next several centuries with the melting of the Greenland ice sheet and another 6 meters with the melting of the West Antarctic ice sheet, James Hansen, head of the climate science program at the NASA's Goddard Institute for Space Studies in New York, suggests those estimates are conservative. He suggests that sea level will rise far more quickly, possibly by as much as 25 meters by 2100. See Hansen, J. (2007). Huge sea level rises are coming unless we act now. *New Scientist*, 2614, 30–4, http://environment.newscientist.com/channel/earth/mg19526141.600-huge-sea-level-rises-are-coming – unless-we-act-now.html
9. Lowe, J. A., Gregory, J. M., Ridley, J., Huybrechts, P., Nicholls, R. J., and Collins, M. (2006). The role of sea-level rise and the Greenland ice sheet in dangerous climate change: Implications for the stabilisation of climate. In *Avoiding Dangerous Climate Change*, ed. H. J. Schellnhuber, W. Cramer, N. Nakicenovic, T. Wigley, and G. Yohe. Cambridge: Cambridge University Press, pp. 29–36.

10. For example, because of its abundance and relative low cost, future coal use is expected to increase under any foreseeable future scenario (this is certainly the case in developing China and is already playing havoc with air pollution in urban areas) and could make future attempts to reduce greenhouse gas emissions exceedingly difficult. However, new clean coal technologies have been developed. For example, thermo-energy integrated power system (TIPS) may reduce emissions. TIPS keeps coal under extreme and constant pressure, stripping it of virtually all its pollutants while generating relatively low-cost electricity. See Interest develops in a new high efficiency alternative clean coal power generation technology for CO_2 capture and sequestration. *Canadian CO_2 Capture and Storage Technology Network Newsletter* (February 2007). See also "Clean coal" on the Natural Resources Canada Web site, http://canmetenergy-canmetenergie.nrcan-rncan.gc.ca/eng/clean_fossils_fuels/clean_coal.html
11. IPCC (2007). Technical summary. In *Climate Change 2007: The Physical Science Basis*, p. 77.

17. DEAD ZONES

1. This comment was made by Dr. Peter Keller, a professor of geography at the University of Victoria, as he summed up the talk "Scientific Uncertainty, Social Conflict and Social Decay during Fishery Crises," given by Professor Renato Quiñones from the Center for Oceanographic Research in the Eastern South Pacific, University of Concepción, Chile (presented at the forum titled "Are We Killing the World's Oceans?" University of Victoria, February 21 and 22, 2007).
2. On the ScienceDaily Web site, www.sciencedaily.com/releases/2007/02/070222155140.htm, from material provided by Cornell University, February 23, 2007.
3. Information from comments made by Dr. Lisa Levin, Scripps Institution of Oceanography, University of California San Diego, during her presentation, "Why Are Dead Zones in the Sea Proliferating?" (presented at the forum titled "Are We Killing the World's Oceans?" University of Victoria, February 21 and 22, 2007).
4. Intergovernmental Panel on Climate Change (IPCC) (2007). *Climate Change 2007: The Physical Science Basis*. Contribution of Working Group I to the Fourth Assessment Report of the IPCC, ed. S. Solomon, D. Qin, M. Manning, Z. Chen, M. Marquis, K. B. Averyt, M. Tignor, and H. L. Miller. Cambridge: Cambridge University Press, pp. 5, 11, available online at http://www.ipcc.ch/publications_and_data/ar4/wg1/en/contents.html
5. "Ocean around Japan warming up fast," Reuters News Service, May 16, 2007, www.planetark.org/dailynewsstory.cfm/newsid/41923/story.htm
6. Based on comments made by Dr. Daniel Pauly, University of British Columbia, during his presentation, "Are We Past the Point of No Return in Mining Fish From the Sea?" (presented at the forum titled "Are We Killing the World's Oceans?" University of Victoria, February 21 and 22, 2007).
7. Quote and comments by Dr. Ken Caldeira, Carnegie Institution, Stanford University, during his talk, "Is Human-Induced Acidification of the Sea About to Destroy Reefs?" (presented at the forum titled "Are We Killing the World's Oceans?" University of Victoria, February 21 and 22, 2007).
8. Foraminifera are single-celled organism with shells that have lived in the oceans for at least 540 million years. There are about 4,000 known living varieties; their

prevalence has changed with changing ecological conditions throughout Earth's history. Many scientists believe they are a key group in the food chain, being an important food source for small marine invertebrates and fish.

9. Fabry, V. J., Seibel, B. A., Feely, R. A. and Orr, J. C. (2008). Impacts of ocean acidification on marine fauna and ecosystem processes. *ICES Journal of Marine Science*, 65, 414–32. See Web site: http://icesjms.oxfordjournals.org/content/65/3/414.full.pdf+html
10. From a talk by Professor John Veron, former Chief Scientist of the Australian Institute of Marine Sciences, to the Royal Society of London, 2009, available online at http://john-blatchford.suite101.com/coral-reef-crisis-royal-society-lecture-a146481
11. Sumaila, U. R., and Pauly, D., eds. (2006). *Catching More Bait: A Bottom-Up Re-Estimation of Global Fisheries Subsidies*. Fisheries Centre Research Reports 14(6). Vancouver: The Fisheries Centre, University of British Columbia. See particularly Khan, A., Sumaila, U. R., Watson, R., Munro, G., and Pauly, D. The nature and magnitude of global non-fuel fisheries subsidies, pp. 5–37.
12. Check the Seafood Watch Web site (www.montereybayaquarium.org/cr/seafoodwatch.asp) for information about sustainable fish products and practices.
13. Based on comments made by Dr. Harald Rosenthal from the University of Kiel, Germany, in his presentation, "Aquaculture, the Environment and Sustainability Issues" (presented at the forum titled "Are We Killing the World's Oceans?" University of Victoria, February 21 and 22, 2007).
14. Robert G. B. Reid, personal communication, June 2007.
15. Based on comments made by Dr. Daniel Pauly, University of British Columbia, during his presentation, "Are We Past the Point of No Return in Mining Fish from the Sea?" (presented at the forum titled "Are We Killing the World's Oceans?" University of Victoria, February 21 and 22, 2007).

18. THE ECONOMIC CONNECTION

1. Macpherson, C. B. (1965). *The Real World of Democracy*. CBC Massey Lecture Series. Scarborough, ON: CBC Enterprises, p. 2.
2. Smith, A. (1910). *The Wealth of Nations*. New York: Random House, p. 399. First published in 1776.
3. Macpherson, *Real World of Democracy*, pp. 12–13.
4. Ibid., p. 13.
5. Ibid., p. 85.
6. Galbraith, J. K. (1958). *The Affluent Society*. Toronto: New American Library of Canada, p. 45.
7. Gray, J. (2003). *Al Qaeda and What It Means to Be Modern*. London: Faber and Faber, p. 48.
8. Ibid, p. 53.
9. Alan Greenspan, chairman of the U.S. Federal Reserve Board from 1987 to 2006, in the BBC documentary *For the Love of Money*. Part 2, *The Age of Risk*.
10. Securitized investment products are the result of a process, securitization, in which a new investment instrument is created by combining various financial assets – e.g., mortgages – into one large pool, repackaging them, and selling them to investors.

11. Greenspan, in *For the Love of Money*. Part 2, *The Age of Risk*.
12. Gray, *Al Qaeda*, p. 14.
13. Ibid, p. 15.
14. Quoted by Bernhut, S. (2003). Leader's edge: An interview with Professor John Kenneth Galbraith. *Ivey Business Journal*, 68, no. 1, reprint #9B03TE04, p. 7, www.iveybusinessjournal.com/view_article.asp?intArticle_ID=436
15. Ibid., p. 1.
16. See USGS Earthquake Hazards Program Web site on the March 11, 2011, Japanese earthquake, http://earthquake.usgs.gov/earthquakes/eqinthenews/2011/usc0001xgp/#summary
17. See USGS Earthquake Hazards Program Web site on the January 12, 2010, Haiti earthquake, http://earthquake.usgs.gov/earthquakes/recenteqsww/Quakes/us2010rja6.php#summary
18. See USGS Earthquake Hazards Program Web site on the December 26, 2004, Northern Sumatra earthquake, http://earthquake.usgs.gov/earthquakes/eqinthenews/2004/us2004slav/#summary
19. Rees, W. E. (1996). The ecology of community sustainability: Global context, local action. Paper presented to the International Summer Institute on Participatory Development, University of Calgary, July 15–19, p. 1. Originally presented to the OECD-Germany Conference on Sustainable Urban Development, Berlin, March 19–21, 1996, under the title "Urban ecological footprints: Biophysical and thermodynamic dimensions of sustainability."
20. Macpherson, *Real World of Democracy*, p. 95.
21. Ibid., p. 87.
22. Ibid., p. 31.
23. Keynes, J. M. (1932). *Essays in Persuasion*. New York: Harcourt Brace, pp. 369–72.

19. THE PROGRESS OF DOMINANCE

1. John Gray discusses the influence and impact of American administrative policy on the world in Gray, J. (2003). *Al Qaeda and What It Means to Be Modern*. London: Faber and Faber, pp. 85–7.
2. For a ten-year account of major foreign holders of U.S. Treasury securities, see the U.S. Department of the Treasury Web site, http://www.treasury.gov/resource-center/data-chart-center/tic/Documents/mfh.txt
3. Gray, *Al Qaeda*, pp. 85–7.
4. See the comments by John Kenneth Galbraith in Bernhut, S. (2003). Leader's edge: An interview with Professor John Kenneth Galbraith. *Ivey Business Journal*, 68, no. 1, reprint #9B03TE04.
5. Publication of the Stern Review on the Economics of Climate Change. Press release from HM Treasury, October 30, 2006, www.hm-treasury.gov.uk/newsroom_and_speeches/press/2006/press_stern_06.cfm
6. Stern, N. (2006). Executive summary. In *Stern Review: The Economics of Climate Change*. London: HM Treasury, p. xvi. Available online at www.hm-treasury.gov.uk/independent_reviews/stern_review_economics_climate_change/stern_review_report.cfm
7. In Diamond, J. (2005). *Collapse: How Societies Choose to Fail or Succeed*. New York: Penguin, Jared Diamond gives examples of corporations that have taken the long view and are already benefiting from more environmentally aware strategies.

8. As restated by Rees, W. E. (1996). The ecology of community sustainability: Global context, local action. Paper presented to the International Summer Institute on Participatory Development, University of Calgary, July 15–19, p. 2. Originally presented to the OECD-Germany Conference on Sustainable Urban Development, Berlin, March 19–21, 1996, under the title "Urban ecological footprints: biophysical and thermodynamic dimensions of sustainability."

20. DANGEROUS ATTITUDES

1. We have been warned. Now everyone should understand why we have to combat climate change, (editorial) *New Scientist*, November 4, 2006, p. 4.
2. For a very thoughtful exposé on the food shortage in Somalia read Keneally, T. (2011). The politics of hunger: part I. *Globe and Mail,* September 3, p. F1, F6–7.
3. Statements made by Dr. Anwar A. Abdullah, Senior Advisor to the Prime Minister on Sustainable Development, Kurdistan Regional Government, at the Climate/Security conference on 9 March 2009 at the University of Copenhagen. First published in Hetherington, R., and Reid, R.G.B. (2010). *The Climate Connection: Climate Change and Modern Human Evolution.* Cambridge: Cambridge University Press.
4. Personal communication from Dr. Anwar A. Abdullah, 6 April 2009. First published in Hetherington and Reid (2010). *The Climate Connection: Climate Change and Modern Human Evolution.* Cambridge: Cambridge University Press, p.286.

21. HELPFUL STRANGERS

1. Thomas, L. (1974). *The Lives of a Cell: Notes of a Biology Watcher.* New York: Viking Press, p. 3.
2. Ibid p. 99.
3. Rees, W. E. (1996). The ecology of community sustainability: Global context, local action. Paper presented to the 1996 International Summer Institute on Participatory Development, University of Calgary, July 15–19, p. 6. Originally presented to the OECD-Germany Conference on Sustainable Urban Development, Berlin, March 19–21, 1996, under the title "Urban ecological footprints: biophysical and thermodynamic dimensions of sustainability."
4. Hoegh-Guldberg, O. et al. (2007). Coral reefs under rapid climate change and ocean acidification. *Science*, 14, 1737–42.
5. The information in this paragraph is drawn from the results of the Census of Marine Life. For a description and links to more information, see the Census Web site at http://www.coml.org/about-census
6. Black, R. (2006). "Only 50 years left for sea fish." BBC News, UK version, November 2, http://news.bbc.co.uk/2/hi/science/nature/6108414.stm
7. Robert G. B. Reid, personal communication, 2007.

22. TRIUMPHANT OBLIVION

1. For a nice article on this, see Martell, A. (2011). The species problem: What do you get when you cross a polar bear with a grizzly? *The Walrus*, April, pp. 17–18, http://www.walrusmagazine.com/articles/2011.04-frontier-the-species-problem/
2. Reid, R. G. B. (2007) *Biological Emergences: Evolution by Natural Experiment*, Vienna Series in Theoretical Biology. Cambridge, MA: MIT Press.

3. See, for example, Paddock, C. (2011). Soil bacteria help kill cancer tumors. *Medical News Today*, September 5, http://www.medicalnewstoday.com/articles/233879.php

23. OUR CHILDREN

1. See Lewis, S. (2005). *Race Against Time*. Toronto: House of Anansi Press, for an extraordinary account of Africa's plight.

24. LIVING IN A DANGEROUS CLIMATE

1. Daniel Quinn (1992) does a wonderful job of portraying these ideas in his book *Ishmael: An Adventure of the Mind and Spirit*. New York: Bantam Books.
2. United Nations Environment Programme (2007). *Global Environment Outlook 4: Environment for Development*. Valletta, Malta: Progress Press.
3. Ibid., p. 459.
4. Kahfeld, H. (1995). Global rise in fuel consumption by the transport sector undermines aim of reducing CO_2 emissions. *Economic Bulletin*, 32, no. 4; see also Commission on Engineering and Technical Systems. (1992). *Automotive Fuel Economy: How Far Can We Go?* Washington DC: National Academies Press, p. 12, http://www.nap.edu/catalog.php?record_id=1806
5. Hetherington, R., and Reid, R. G. B. (2010). *The Climate Connection: Climate Change and Human Evolution*. Cambridge: Cambridge University Press, p. 294.
6. Personal communication with Dr. Ursula Franklin, April 16, 2008. First published in Hetherington and Reid (2010). *The Climate Connection: Climate Change and Modern Human Evolution*. Cambridge: Cambridge University Press, p. 286.

Index

Abdullah, Anwar A., 177, 244
Abu Hureyra, 82, 209
Acheulean stone tool technology, 28, 30, 32, 53, 209, 212, 229, 231
Adams, Jonathan, 231
adaptability, i, x, xvi, 6, 17, 96, 99–101, 111, 185, 209
adaptation, x, xvi, 17–19, 27, 96, 99–101, 143, 178–9, 201, 209, 238, 240
Africa, xv, 3, 7–9, 11, 23–30, 32, 34, 39, 42, 48, 50–7, 67–70, 78–80, 82–5, 94, 104, 113, 125, 134–5, 137–8, 140, 143, 185, 192, 201, 206, 209, 211, 213, 217–18, 221, 223, 228–9, 231, 233, 237, 239, 245
agricultural belt, 16
agriculturalists, 11–12, 77, 101–2, 209
agriculture, i, v, xv, 11–12, 18, 70, 73–4, 76, 79, 81–2, 85, 87, 91–6, 101, 103, 137–8, 140, 149, 152, 196, 200, 209, 214, 219, 222, 236–7. *See also* domestication
Akkadian Empire, 79, 84
Alaska, 58, 65, 68–9, 215, 233–5
albedo, 209
Alero Tres Arroyos, Tierra del Fuego, 63
Allchin, B., and Allchin, R., 236
Altai Mountains, Siberia, 67
Amazon rainforest, 140
Ambrose, S.H., 229
American lion, 61
Americas, xv, 10, 39–40, 43, 58, 61–9, 94, 140, 143, 156, 221, 233–5
Amerindian, 67, 235
amplification, 116–17
anatomically modern humans, 29
Andersson, Lief, 236
Antarctic ice sheet, 143
Antarctica, 9–10, 13, 48–50, 54, 219, 228, 232, 240
aquaculture, 152
archaic *Homo sapiens*, 56
Arctic, 4, 58, 77–8, 135, 140, 148, 185, 202, 215, 227–8
Ardipithecus ramidus, 27
aridity, 16, 237
Aristotle, 24
Arnell, N., and Liu, C., 240
art, 29, 38, 57, 95, 210, 213
Asfaw, B. et al., 229
Asia, 3, 7, 10, 11, 28–30, 42, 48, 50–1, 54, 56, 59, 64, 66, 68–70, 77–81, 134, 138, 140, 218, 225, 228–9, 232, 235–6
Assyrian Empire, 79, 216
Atlantic meridional overturning, 136–8, 143, 210
Atlantic Ocean, 11, 66, 83, 93, 136, 138, 140, 148, 210
Aurignacian stone tool technology, 37, 210
Australia, 3, 10, 16, 40, 43, 48, 55–6, 76, 104, 135, 137, 140, 143, 201, 232
Australopithecus, 7, 25–9, 42, 211, 216, 224
Australopithecus aethiopicus, 27, 211
Australopithecus afarensis, 25–7, 211
Australopithecus africanus, 27–8, 211
Australopithecus anamensis, 27, 211
Australopithecus robustus, 27, 211, 224
Australopithecus sediba, 28–9, 211
Awash Valley, Ethiopia, 27

Babylonian Empire, 79
bacteria, 11, 100, 128–9, 180, 186, 197, 245
Baiji dolphin, 135
Baker, P.A. et al., 237

Bakola Pygmies, 125
bananas, 79–80
Bangladesh, 140
Banks, W.E. et al., 230
Barber, D.C. et al., 236
barley, 79–81, 86, 209
Bar-Yosef, O., and Belfer-Cohen, A., 231
Bateson, William, 120–1
beans, 79–80
Beatles
  Abbey Road, 116
Beckett, Margaret, 177
behavior
  behavioral adjustments, 39
  behavioral revolution, 43, 55, 57, 101
  symbolic behavior, 29, 38–9
Belyaev, D.K., 236
Berger, Lee, 28, 211
Beringia, 9, 59–61, 63–4, 68, 211, 213, 215, 225, 233
Beringian Paleo-Arctic people, 62
Bernanke, Ben, 162
Bernhut, Stephen, 155
Berry, Wendell, 184
bifacial tools, 37, 61, 212
big bang of animal evolution, 4
biodiversity, 19, 181–2
bipedal apes, 7, 25, 217, 227
Bipindi, South Camaroon, 125
bison, 61
black bear, 65
Black Sea, 83
Black, R., 244
Blackburn, T.C., 234
Bluefish Caves, Yukon, 64
bolide impacts, 4, 6, 12
Bonnichsen, R., and Steele, D.G., 234
Bose basin, China, 30, 229
bottle gourd, 69
Bowler, J.M. et al., 232
Boxgrove, England, 56
Brahic, C., 240
brain size. *See* cranial capacity
Brazil, 16, 63, 66, 235
breadfruit, 79–80
Brinkhuis, H. et al., 227
Britain, 157, 219
British Columbia, i, 64, 150, 233–4, 241–2
British Empire, 12, 168
Britten, R.J., 229
Broca, Paul, 42, 231
Brooks, N., 237
Broom, Robert, 27
brow ridges, 29, 36, 213, 218
Brumm, A. et al, 230
Buffon, George Louis Leclerc, 24
Burroughs, W.J., 80, 237
Butler, S., 228
bycatch, 151

C3 grasses, 56
C4 grasses, 56
Caldeira, Ken, 150, 241
California, 16, 65, 67, 69, 123, 234–5, 241
Callaway, E., 230
camels, 78
Canada, i, xiii, 15, 59, 63–5, 68, 78, 81, 93, 112, 135, 140, 146–8, 160, 163, 182, 192, 198, 211, 227, 233–5, 241–2
Canadian Rocky Mountains, 61, 63
Capelinha, Brazil, 66
capitalist market system, 156–7, 159–60, 165–6, 183
  capitalist market society, 156–7
  free market, 107, 155, 157–8, 162, 165, 198
Caral, Peru, 94–5
carbon dioxide, 4, 10, 13–17, 20, 75, 133–4, 136–40, 143–4, 150–1, 202–3, 215–16, 224–5, 228, 240–1, 245
Cariaco Basin, 91
cats, 6, 78, 80
cattle, 78, 80, 84, 237
  auroch, 78
  ox, 78, 95
Cayapa tribe, Ecuador, 67
Census of Marine Life, 82, 182, 244
Central America, 62
Chakravarti, A., 229
Châtelperronian stone tool technology, 37–8, 210, 212
Chauvet cave, France, 210
Chen, C. et al., 228
chickens, 78
chickpeas, 79
Chicxulub comet, 5
Childe, V.G., 73, 222
  oasis theory, 74, 222
Chile, 63, 68, 92, 241
chilli peppers, 79–80
chimpanzees, 25, 28, 34, 216, 223–4, 229
China, 7, 11, 16, 28–31, 39, 69–70, 78–81, 135, 140, 143, 152, 169, 218, 227, 229, 241
chromosomes, 67, 124, 212–13, 217, 225–6, 235
Chumash tribe, 67–8
Churchill, Stephen, 29
Cinq-Mars, Jacques, 64
civilization, xv, 42, 74, 81, 85, 87, 92, 94–6, 200, 216, 235–7
Clark, J.D., 229

climate catastrophe, 3–4, 6, 19, 87, 100, 103–4, 114, 176, 182
  rapid climate change, xvi, 13, 16, 18–19, 35, 40, 44, 47, 96, 105, 109, 113, 135, 171–2, 178, 197, 244
climate variability, 16
climatic instability, 18, 51–2, 103
climatic stability, 12, 101, 114
  stable climate, 12, 86, 101, 171, 197
Clovis, New Mexico, 61, 212
Clovis culture, 63
Clovis first hypothesis, 61–2, 64
Clovis people, 61, 63, 212, 223
Clovis stone tools, 61–3, 66, 212
coastal migration route, 65–6, 68
Cobá, Yucatan Peninsula, 88–9
coconuts, 69, 79–80
Colombia, 67
Columbia River, Washington, 62, 66, 220
competition, 12, 39, 100–1, 105–10, 113, 115, 144, 155, 157–9, 163, 165–6, 177, 196, 201, 203, 213
consumption, 18, 20, 105, 133, 164, 181, 183, 205, 245
continental shelves, 9–10, 49–51, 54, 56, 58–60, 65, 76, 149, 211, 225, 231, 233–4
coral reefs, 137, 140, 150–1, 182
Cordilleran ice sheet, 63
corn, 69, 76, 79–80, 88, 106. *See also* maize
cotton, 69, 79–80, 93, 236
cradle of civilization, 79. *See also* Mesopotamia; Fertile Crescent
cradle of humankind, 24
cranial capacity, 27–9, 32, 37, 42, 213, 217–18
Crawford, M.A. et al., 231
creative spark, 24, 31, 34, 57
Cretaceous period, 4, 6
  Cretaceous extinction, 5, 151
crisis, communication, and collaboration, 19, 51, 87, 200, 202
  3 Cs, 19, 94, 202, 228
Cro-Magnon, 42–3, 213
cultural renaissance, 39. *See also* creative spark: behavioral revolution
Cuvier, Baron Georges, 6

Dalton, R., 235
Dart, Raymond, 27
Darwin, Charles, v, x, xv, 17, 23–5, 105–8, 110, 112–15, 117–19, 121, 123–6, 168, 178, 197–8, 213, 228, 238–9
  Darwinian theory of evolution, 17–18, 213
  The Descent of Man, and Selection in Relation to Sex, 228
  On the Origin of Species, 23, 114, 123, 238–9
Dauenhauer, N.M. and Dauenhauer, R., 233
Davis, N.Y., 235
Dawkins, Richard, 118
dead zones, 149, 151
democracy, xiv, 155, 163, 166, 170, 198, 242–3
  democratic franchise, 156
  liberal-democracies, 155, 167
dengue fever, 143
Dennell, R.W., 231
Dennell, R.W. et al., 228
dentition, 31, 213
  teeth, 29, 62, 213, 218
deoxyribonucleic acid
  DNA, 34, 39, 53–4, 67, 116, 124, 128–30, 180, 185, 212–14, 216, 221–2, 229, 232, 235, 239
desert, 9, 47, 50–2, 56, 81, 84–5, 88, 106, 228
developmental epigenetics, 185
Di Lernia, S., 237
Diamond, J., 80, 236–7, 243
Dikov, Nikolai, 62
Dillehay, T.D. et al., 236
dire wolf, 61
disease, 50, 82, 102, 128–9, 140, 143, 215, 219
dispersal, 32, 213, 229, 234
diversity, v, x, 25, 41, 65, 106, 150, 162, 170, 182, 193, 197–8, 203–6
Dixon, E. James, 65, 234
Dmanisi, Georgia, 228–9
dogs, 6, 77, 80–1, 117
domesticate plants and animals, 11
  domestication, i, 11, 70, 77–81, 87, 93, 95, 123, 125, 214, 236
dominant attitude, x, 170, 179, 183
  right to dominate, 40
dominant societies, i, 20, 102, 105, 108, 113, 166–71, 193
dominant species, xvi, 6, 11–12, 18–19, 100, 105, 108–9, 114, 171, 178, 182–4, 196–7, 199, 214
donkeys, 78, 80
Dow, G.K. et al., 237
Down syndrome, 124
Dromaius novaehollandiae, 56
drought, 3, 12, 16, 47, 52, 73–4, 76–7, 84–6, 91–2, 94, 104, 126, 135, 137, 140, 200–1
  drought-ridden, 3, 16
Dryanda forest, Australia, 137, 140

Dubois, Eugène, 28
dwarfism, 42, 125, 231
dwarfs, 42
Dyke, A.S., 233
dynamic stasis, 6, 101, 109, 214, 238
  dynamic equilibrium, 18

Easter Island, 193–4
Eby, M. et al., 240
Eby, Mike, 114
economic instability, 18, 102, 171–2, 177
  depression, 158, 164, 166
  recession, 3, 158, 162, 177
economic stability, 12, 171
economic systems, ix, 11
ecosystems, 18, 137–40, 144, 148, 178, 181, 214
Ecuador, 67, 69
Egypt, 82, 85, 94
Egyptian empire, 85
Ehlers, J., and Gibbard, P.L., 48, 233
einkorn wheat, 79–80, 82, 209
Einstein, Albert, 36, 145, 165, 175, 178
El Mirador, Yucatan Peninsula, 91
El Niño-Southern Oscillation, 75–6, 93–4, 104, 214, 234–7
Eldredge, N., 238
Eldredge, Niles and Gould, Stephen Jay, 114–15, 238
elephant, 41, 43, 54, 78, 214
Ellis, C. et al., 230
emergence, 18–19, 23–4, 51, 83, 94, 114, 185, 214
Emergence theory, 18–19, 214
emissions, 3, 13–14, 133–4, 136–7, 139, 143–4, 228, 241, 245
  emission-reduction, 3
emmer wheat, 79–80, 209
Energy Policy and Conservation Act, 81, 205
England, 56, 66, 120, 156, 235
environmental instability, 18, 201
environmental stability, 6, 100, 109, 162, 201
Eocene epoch, 6
Eritrea, 51
Eskimo-Aleut, 62, 67, 215
Estés, C.P., 238
Ethiopia, 27, 32, 50, 104, 229
Ethiopian River Valley, 25
Eurasia, xv, 9, 32, 49, 52, 54, 81, 231
Europe, 3, 7, 9–11, 15, 29–30, 32, 34, 36–9, 43, 48, 50–1, 54–7, 66, 69, 78–9, 81, 134, 136–8, 143, 163, 209–10, 212–13, 218, 221, 230, 234, 236–7
Euxine Lake, 83. *See also* Black Sea

evolution, gradual genetic change, xii, 17, 19, 100–1, 105, 114, 122–3, 144, 201, 216, 225, *See also* Darwinian theory of evolution
evolution, rapid, 18, 52, 123, 127, 130, 178. *See also* Emergence theory: saltation: saltatory
evolution, social, 19, 225
extinction, i, xvi, 4, 6–7, 10, 18–19, 29, 31, 34–5, 38–44, 54–6, 59, 61, 69, 96, 100, 105, 108, 135, 137, 140, 143, 178, 193, 197, 199, 201, 203, 210, 214, 217–18, 220, 223–4, 230–1, 233
  mass extinctions, 4, 6, 151–2, 182
extreme climate events, 16, 109, 138, 201

Fabry, V.J. et al., 242
Fagan, B., 80, 236–7
famine, 104, 127–9, 135, 177, 205, 239
farmers, x, 3, 11, 73, 76–9, 81–4, 88, 101, 124
Faure, H. et al., 231
Felis sylvestris, 78
Ferretti, G. et al., 239
Fertile Crescent, 79–83, 92, 209, 216
Finlayson, C., 230
fire, 17, 28–31, 99–100, 135, 217–18
Fladmark, Knut, 65, 233
Flannery, T., 230
Flinders, Carol Lee, 41, 99, 155
floods, 74, 85, 93, 103, 164, 201
  flooding, 3, 79, 83, 85, 103, 138–40
Flores Island, Indonesia, 28, 41–3, 54, 214, 217–18, 220, 230–2
Florida, 16, 67
Foley, R.A., 231
Folsom, New Mexico, 61, 212
Fondon, J.W., III, and Garner, H.R., 238
foraminifera, 241
Forster, P., 53, 232
Forster, P., and Matsumura, S., 53
fossil fuels, i, 12–13, 15, 18, 95, 102, 105, 134, 144, 169, 175–6, 178, 186, 193, 205, 216
fossils, x, 25–6, 29–32, 34, 37, 39, 42, 50, 52, 57, 114, 216–17, 223, 227–8, 230, 235, 241
foxes, 77
France, 38, 156, 209–10, 212–13
Franklin, U.M., 207, 228, 245
  The Ursula Franklin Reader, Pacifism as a Map, 228

Gabunia, L., Antón, S.C. et al., 229
Gabunia, L., Vekua, A. et al., 228
Galbraith, John Kenneth, 155, 163, 243
Galton, Francis, 121
Garcia-Cao, I. et al., 239
Gathorne-Hardy, F.J., and Harcourt-Smith, W.E.H., 232
genes, 34, 39, 100, 109, 112–13, 117–18, 124, 126–30, 185–6, 197–8, 213, 216–17, 221–2, 224, 238
genetic, i, 17–18, 34, 39, 67–8, 77–8, 96, 99–101, 105–6, 108, 113–17, 122–6, 129, 185–6, 213, 216, 219, 222, 229, 235, 238
  genetic similarities, 39
  genetic variations, 34
genome, 39, 124, 129, 216–17, 222, 229–30, 232
genomists, 34, 216
Genyornis newtoni, 56, 230, 233
Georgia, 30, 228
germ line, 123, 216
Germany, 31, 38, 57, 121, 123, 242–4
giantism, 125
Gibbons, A., 229
Gilmore, D.Y. and McElroy, L.S., 235
Giuffra, E. et al., 236
glacial cycles, 19
  glacial interval, 54–5, 57
glacial lake, 10, 59, 64, 93
glaciations, 9, 233
glaciers, 9–10, 13, 47, 49, 58, 137
  ice sheets, 8–9, 13, 15, 49, 52, 54, 56, 58–9, 61, 63–5, 75–6, 81, 126, 136–7, 139, 219
global average sea level, 9–10, 41, 43, 47, 49–54, 56, 59, 76, 135, 139–40, 211, 227, 236–7, 240
global economic crisis, 169
global financial crisis, 159, 166, 169, 171
global warming, ix, 3, 74, 126, 133, 137–9, 144, 152
goats, 78, 80–2, 86
golden toad, 135
Goldschmidt, Richard, x, 107, 123–5, 197, 218
  hopeful monsters, 107, 110–11, 122, 124–6, 197, 218
gorillas, 25, 216
Goswami, B.N. et al., 240
Gould, S.J., 231, 238
Gould, S.J., and Eldredge, N., 238
Grand Banks, Canada, 148, 182
Gray, Elizabeth Dodson, 238
Gray, John, 158–60, 162, 166, 168–9, 243
  Al Qaeda and What it Means to be Modern, 158, 243
great apes, 216–17, 227. *See also* hominid
Great Karoo grass flats, South Africa, 137, 140
Green, R.E. et al., 230
Greenfield, P.M., 229
greenhouse gases, 3, 13, 75, 113, 126, 133–4, 136, 139, 216
  greenhouse gas emissions, 16, 20, 241
greenhouse warming effect, 4, 13, 17
Greenland, 9, 66, 68, 136, 148, 215, 219
Greenland ice sheet, 139–40, 240
Greenspan, Alan, 159, 162, 165, 242–3
Greenwood, E., 239
Grijalva River delta, Mexico, 76, 79
grizzly bears, 58
Gruhn, Ruth, 65, 234
guinea fowl, 79–80
Gulf of Mexico, 65, 149
Guthrie, Arlo, 116, 118
Guthrie, Arlo and Goodman, Steve
  City of New Orleans, 116, 206
Guthrie, R.D., 233
Guthrie, Woody, 116–18
  The Farm-Labor Train, 116
  This Land Is Our Land, 116, 206

Hadar, Ethiopia, 27
Haeckel, Ernst, 121
Haida Gwaii, i, 65
Haiti, 103, 164, 201, 243
hand axes, 28, 30, 53, 212, 221
Hansen, J., 240
Harappan civilization, 86
Haug, G.H. et al., 237
heat stress, 6, 128
Heaton, T.H., Talbot, S.L. and Shields, G.F., 234
heat-shock genes, 129
heat-shock proteins, 128–9, 217, 239
helium, 4
Hetherington, R. and Reid, R.G.B., 48, 53, 60, 80, 143, 227–34, 236–7, 244–5
  The Climate Connection, i, xiii, 48, 53, 60, 80, 143, 227–30, 232–3, 236, 244–5
Hetherington, R., Barrie, J.V., MacLeod, R., and Wilson, M., 227, 233–4
Hetherington, R., Barrie, J.V., Reid, R.G.B. et al., 234

Hetherington, R., Wiebe, E. et al., 231, 233
Heusser, C.J., 65, 233
Hittite empire, 85
Hodell, D.A. et al., 237
Hoegh-Guldberg, O. et al., 244
Holden, C., 234
Hole, F., 237
Holocene, 12, 70, 81, 85–6, 92, 101, 115, 230, 234–7
hominid, 25, 29, 216–17, 227–8, 231, 234
hominin, 7, 25–32, 42–3, 210–11, 217–18, 224, 227–9, 231
Homo, i, v, xv, xvi, 7, 10, 18–20, 23, 25–32, 34–5, 37–8, 40–2, 44, 47, 54, 56, 69, 101, 105–6, 108, 113, 115, 171, 178, 181, 183, 195, 197, 200, 205–6, 209–10, 212–14, 217–18, 220–1, 223, 226, 229, 231, 235
Homo antecessor, 7, 31, 217
Homo erectus, 7, 28–32, 42–3, 47, 56, 69, 200, 209, 217–18, 229
Homo ergaster, 7, 28–9, 31–2, 200, 217
Homo floresiensis, 10, 28, 40, 42–4, 54, 115, 200, 214, 217, 220, 231
Homo habilis, 25–6, 28, 30, 42–3, 200, 218, 223
Homo heidelbergensis, 29–31, 38, 56, 200, 209, 218
Homo neanderthalensis, 29, 35, 37, 200, 218, 231
  Neanderthals, v, 7, 10, 29, 32–3, 36–40, 43, 54–7, 115, 197, 205, 212, 217–18, 221, 230
Homo sapiens, i, xvi, 7–9, 18–20, 28–9, 34–5, 37–40, 42–4, 47, 50–2, 54–7, 69, 101, 103, 105–6, 108, 110, 113, 115, 171, 178, 181, 183, 195, 200, 202, 205–6, 210, 212–13, 217–18, 226, 230–1
Hopkins, M., 233
horses, 6, 24, 31, 38, 61, 64, 78–80, 95
Hou, Y. et al., 229
Hsian, S. et al., 104
Hsiang, S.M. et al., 238
human population, 11, 16, 55, 74, 81–3, 85, 95–6, 102–3, 139, 143–4, 164, 171, 181
hunter-gatherers, 12, 41, 73, 81–2, 99, 102, 228
hunting and gathering, 10, 73–4, 82–3
Huntington's disease, 116–17, 129, 219
Hurtado de Mendoza, D., and Braginski, R., 235
Huxley, Thomas Henry, 24, 121, 228
hydrogen, 4, 220, 223

Iberian Peninsula, 66
ice age, 9, 19, 75, 113, 134, 219
  last ice age, 9–10, 13, 41, 50, 58–9, 61, 64–8, 76, 81, 99, 211, 220, 225, 227, 234
ice cores, 10, 13, 15, 219
  Dome C, 15
  Taylor Dome, 15
  Vostok, *15*, 228
Iceland, 66, 68, 76, 164
Idaho, 64
India, 6, 69, 78, 80–1, 86, 135, 240
Indian Ocean, 53, 55, 149
Indonesia, 10, 28, 41, 54, 218, 220, 229–32
Indonesian Orang Pendek, 43, 219, 231
Indus basin, 81, 86–7, 94
Industrial Revolution, 12–14, 16, 95, 102, 113, 134, 219
inheritance of acquired characteristics, x, 119–20, 122
innovation, 199–205, 221
Institute for Experimental Biology, 119–20
interglacials, 19, 219
Intergovernmental Panel on Climate Change, 82, 143, 239–41
interstadials, 49, 219
intertropical convergence zone, 84–5, 219
  ITCZ, 84, 219
Inuit, 77, 135, 140, 215
Iran, 79
Iraq, 79, 113
Ireland, 66
irrigation, 11, 83–5, 87, 90, 95, 102
Israel, 7, 34, 52, 55, 220
Italy, 7, 31, 217

Jablonka, E., 239
Jablonski, N.G., 234–5
Japan, 62, 68–9, 103, 105, 123, 150, 164, 201, 220, 241
Java, 7, 30, 41
Jeffrey, S., 239
Jericho Springs, 82
Johanson, Donald, 27
Jomon, Japan, 62, 69, 235
Jomon-Ainu, 62, 220
Joplin, Janis
  Pearl, 116
Jousse, H., 237
jujube, 81, 86
Jurassic Mother, 6, 227

Kahfeld, H., 245
Kaiser Wilhelm Institute for Biology, 123
Kakadu wetlands, 140
Kalahari, 137, 140
Kamchatka Peninsula, 68
Kamchatka River, Siberia, 62
Kammerer, Paul, v, 118–23, 185, 197
Kaplan, J.O., 233
Kazakhstan, 79–80
Kebara 2 Israel, 55
Keefer, D.K., deFrance, S.D., et al., 234
Keller, G. et al., 227
Keller, Peter, 146, 241
Kemp, B.M. et al., 235
Kempler, D., 229
Keneally, T., 244
Kennewick Man, 62, 66, 220
Kennewick, Washington, 62
Kenya, 104
Keynes, John Maynard, 158, 167
Khan, A. et al., 242
Khuzestan, 79
Klasies River Mouth, South Africa, 51
Klein, R.G. et al., 231
Komodo dragons, 28, 41, 43, 54, 220
Krieger, A.D., 65, 233
Kromdraai cave, South Africa, 27, 211

ladder of life, 24
Laetoli, Tanzania, 27
Lagoa Santa, Brazil, 66
Lahr, M.M., 235
Lake Agassiz, 11, 59, 64, 81
Lake Bonneville, 59
Lake Chichancanab, 90
Lake Lahontan, 59
Lake Lunkaransar, India, 86
Lake McConnell, 64
Lake Missoula, 64
Lake Titicaca, Peru, 93
Lake Turkana, Kenya, 7, 27–8, 31, 217
Lake Victoria, East Africa, 51–2
Lamarck, Jean-Baptiste, 119–20, 122
Histoire naturelle des animaux sans vertebres, 120
Lambeck, K. et al., 49
language, xi, xv, 18, 23, 27, 32, 34, 38, 65, 95, 110, 211, 222, 229
Lapa Vermelha IV, South America, 67, 234
Larick, R. et al., 229
last glacial cycle, 9, 47, 49–50, 55–6
Late Stone Age, 52, 55, 57
Laurentide ice sheet, 59, 63–4, 83
Leakey, Louis, 28
Leakey, Mary, 27–8
Lehman Brothers Investment Bank, 159, 161–2
lentils, 79–80
Levallois, 32, 220–1
Levant, 7, 52, 85, 220
Levin, Lisa, 241
Lewis, S., 245
linear progression, 24–5
llamas, 78
London, England, 3, 53, 157, 228, 232, 242–3
Lower Palaeolithic, 223
Lumey, L.H., 128, 239
Lüthi, D. et al., 10
Lynas, M., 143

Macaulay, V. et al., 232
Macgowan, K. and Hester, J.A., Jr., 65, 233
Mackenzie River, 59, 64, 211
MacPhee, R.D., and Marx, P.A., 231
Macpherson, C.B., 155, 157, 165–7, 242–3
Maggio, R., 238
maize, 69, 76, 79
Malapa caves, South Africa, 28
malaria, 140, 143
Malthus, Thomas Robert, 108
Essay on the Principle of Population, 108
mammoth, 37, 59, 61, 64, 99, 233
manioc, 79–80, 93
Manitoba, 64
Marine Isotope Stage 2, 47
Marine Isotope Stage 3, 49, 230
Marine Isotope Stage 4, 47
Marine Isotope Stage 5, 49
Marine Isotope Stage 6, 47–9
marine resources, 11, 29, 51
Marland, G. et al., 14, 228
Marshall, Sean, 114
marsupials, 6, 39, 55, 220
Martell, A., 244
Martin, P.S., 230
mastodons, 61
Mauer, Germany, 56
Maya civilization, 12, 88, 91, 96, 200
Maya culture, 88, 91, 168
McBrearty, S., and Brooks, A.S., 232–3
McCarthy, J.J. et al., 240
McKie, R., 231
Meadowcroft, Pennsylvania, 63
Mediterranean climate, 4
megafauna, 39, 55, 61, 220, 230, 233
Meggers, Betty, 69, 235
Mellars, P., 230
Mercier, N. et al., 230
Mesoamerica, 11, 69–70, 79–80, 87, 92, 221, 235–6
Mesopotamia, 78–9, 82–3, 86–7, 94, 178

Mesozoic era, 20
meteorite, 30
methane, 4, 13, 75, 134, 216
methylation patterns, 128–30, 217, 221, 239
Mexico, iv, 67, 76, 79, 88–9, 221, 227, 234
Micronesia, 62
Middle Awash, Ethiopia, 32, 229
Middle East, 3, 39, 52, 56, 78, 216
Middle Palaeolithic, 32, 38, 53, 55, 220–1, 223, 226
Middle Stone Age, 32, 52, 57, 221
migration, v, xv, 7, 11–10, 16, 29, 31, 42, 44, 47, 50–4, 56–9, 61–70, 73, 76, 81, 102, 135, 176, 199, 207, 212, 218, 223, 227, 231, 233, 235–7
Milankovic cycles, 74–5, 221
Milankovic, Milutin, 74, 221, 235
Miller, G.H. et al., 230, 233
millet, 79–80
Mississippi River, 149
mitochondria, 54, 180, 214, 221
mitochondrial DNA, 54, 67, 221
molluscs, 29
monsoons, 50, 76, 83–4, 126
Montana, 64
Monte Verde, Chile, 63
Montenegro, A. et al., 68, 234–5
Morwood, M.J. et al., 230
Mount Eyjafjallajokul, Iceland, 76
Mount Hekla, Iceland, 76
Mount Toba, Sumatra, 54
Mousterian stone tool technology, 29, 37, 221
Movius line, 53, 221
Movius, H.L., 53
Mumba, Tanzania, 51
music, 29, 37, 106, 116, 119, 197
mutations, x, 17, 124–5, 206, 222

Na-Dené, 62, 67, 222–3
Naish, Constance, 233
natural selection, ix, x, xv, xvi, 17–19, 96, 99–101, 105–10, 113–15, 117–19, 121–3, 130, 144, 157, 165, 168, 171, 184–6, 192, 196–8, 201, 206, 213, 215, 222
neocortex, 34, 222, 229
neo-Darwinists, x, 117–19, 123, 127, 206
neo-Lamarckists, 120–1
Netherlands, 127
Neves, W.A., Powell, J.F. and Ozolins, E.G., 234–5
Neves, W.A., Prous, A. et al., 235
Neves, Walter et al., 66
New Guinea, 39, 230, 232
New World, v, 43, 57–8, 145, 233–5
Nile River, 84–5
nitrogen, 134, 149
nitrous oxide, 134, 216
Noble, G.K., 121
Nohoch-Mul, 88, 90–1
North America, 9, 11, 48, 50, 58–9, 62–3, 65–9, 76, 81, 83, 134, 137, 140, 213, 222–3, 225, 230, 233–4
Nubian desert, 85
nucleotides, 34, 222, 229
Nuu-chah-nulth, 68

oases, 50, 78, 84–5, 222
oats, 79–80
ocean acidity, 150–1, 182, 242, 244
ocean currents, 76, 126, 149
ocean productivity, 134
O'Connell, J.F., and Allen, J., 232–3
oil palms, 79–80
Oldowan stone tool technology, 26, 28, 212, 218, 223
Olduvai Gorge, Tanzania, 26, 28, 223
olives, 79–80
Omo, Ethiopia, 50
O'Neill, Dennis, 29
Ontario, i, 64
Oppenheimer, S., 53
Oregon, 149, 234
Orr, P.C., 234
overkill, 39, 223
oxygen, 4, 125, 133–4, 149–50, 183, 223–4
oxygen-minimum zones, 149
Ozette, Washington, 69
ozone, 4, 109, 223

Pääbo, S., 229
Pacific Northwest coast of North America, 3, 11, 59, 65, 233
Pakistan, 30, 103, 164, 201, 228
Paleo-Arctic people, 223
Paleoindians, 62, 66, 213, 223
Pan paniscus, 34, 223
Papua New Guinea, 39
Paranthropus, 7, 27, 34, 211, 224
Paranthropus robustus. *See Australopithecus robustus*
Parry, M.L. et al., 143, 240
pastoralists, 3, 77–8, 81, 102, 224
Patagonia, 67
Pauly, Daniel, 150, 152, 241–2
Pavlov Institute, 121
peanuts, 79–80, 93, 236
Peru, 65, 76, 79–80, 92–4, 137, 140, 214, 234–7
Petit, J.R. et al., 10, 228, 232
photosynthesis, 4, 224

phyla, 4, 224
Picea sp., 31
Pierson, L.J., and Moriarty, J.R., 235
pigs, 78, 80
Pinnacle Point, South Africa, 51
Pirkkala, L. et al., 239
Pithecanthropus erectus. See Homo erectus
placental animals, 6, 224
Pleistocene, 43, 228–35
polar bear, 135, 140, 185, 197, 244
polar desert, 9, 49, 52, 54, 56, 59, 81, 86
Polynesia, 62, 68
Pope, K.O. et al., 236
Portugal, 9, 218
potato, 69, 79–80
Potts, Richard, 24, 228
prehistoric transoceanic crossings, 66, 68
Prince of Wales Island, Alaska, 65, 67
Przibram, Hans, 119, 121
pygmyism, 41, 125, 185, 239

Quebrada Tacahuay, Peru, 65, 234
Queensland rainforest, Australia, 137, 140
Quinn, Daniel, 196, 245
Quiñones, Renato, 241

radial core technology, 32, 224
Rampino, M.R. et al, 227
Rampino, M.R., and Stothers, R.B., 227
Red Sea, 51
Rees, W.E., 181, 244
Reid, Robert G.B., i, v, xi, xiii, xiv, 48, 51, 53, 60, 80, 106, 114–16, 127, 185, 215, 227–30, 232–4, 237–8, 242, 244–5
Biological Emergences, Evolution by Natural Experiment, 106, 114, 116, 127, 185, 215, 227, 238, 244
Evolutionary Theory, The Unfinished Synthesis, 114
Rendell, H.M. et al., 228
ribonucleic acid
RNA, 180, 213, 222, 224
rice, 79–80, 106, 143
Roberts, R.G. et al., 230, 232
Rogers, R.A., 65, 234
Rogers, R.A., Martin, L.D. and Nicklas, T.D., 234
N. Rolland and S. Crockford, 231
Roman Empire, 12, 168
Romania, 57
Rosenthal, Harald, 242
Russia, 121, 137, 140, 143
Ryan, W., and Pitman, W., 236
rye, 82, 209

Sachidanandam, R. et al., 229
sagittal crest, 27, 211
Sahara, 16, 78–80, 84–5, 135, 140, 237
Sahel, 78–80, 135
Saint-Césaire, France, 38, 212, 230
salmon, 11, 147, 152
saltations, 124
saltatory, 18–19, 125, 197, 214, 225. *See also* Emergence theory
Santa Rosa Island, California, 65, 234
Saskatchewan, 64
Scandinavia, 68
Schellnhuber, H.J. et al., 143, 240
Schöningen, Germany, 31, 38
scrapers, 32, 212, 221
sea-surface temperatures, 136, 150
securitization, 160–1, 242
selfish genes, 118
sequential non-repetitive fine motor control, 32, 225
sewn-plank canoe. *See* tomol
sexual selection, 117–18
Shaadaax', 58, 233
sheep, 40, 77–8, 80–2, 86, 146
short-faced bear, 61
Siberia, 59, 62, 67, 211, 215, 222–3
Siegenthaler, U. et al., 228
Sierra de Atapuerca, Spain, 31
silkworms, 79–80
Silva, A.J., and White, R., 239
Smith, Adam, 106–7, 156–7, 165, 168
The Wealth of Nations, 156–7, 242
Smith, John Maynard, 118
Snowball Earth, 4
solar radiation, 4, 12, 74, 137–8, 209, 221
Sollars, Vincent et al., 128, 239
Solomon, S. et al., 143, 240–1
Solutrean stone-tool technology, 66
Somalia, 104, 177, 244
sorghum, 79–80
South America, 11, 48, 62–3, 65–8, 70, 79–80, 87, 92–3
Southeast Asia, 30, 54–6, 62, 79–81, 104
Spain, 7, 31, 38, 213, 217
spears, 29, 31–2, 37–8, 43
speech, 32, 42, 222, 229
squash, 79–80, 93, 236
St. Acheul, France, 209
stadials, 47, 49, 225
Stanford, D. and Bradley, B., 66, 234
Steele, D.G., and Powell, J.F., 235
Stegodon, 41, 43, 54, 214, 231
Sterkontein, South Africa, 28
Stern, Nicholas, 168, 171, 177, 243

stone tools, 24–5, 28–30, 34, 36–7, 42–3, 55, 57, 62, 200, 210, 212, 216, 218, 220–1, 223, 226, 230
Story, Gillian, 233
Straus, L.G., 66, 234
Sumaila, U.R., and Pauly, D., 242
Sumerian Empire, 79, 84, 216
Sundabans wetland, Bangladesh, 140
supermice, 126, 239
surface-air temperatures, 13, 16, 19, 49, 52, 134–5, 137–40, 144
survival of the fittest, i, ix, xiv, xv, xvi, 19, 42, 96, 101, 105–6, 108–10, 113, 115, 118, 122, 130, 144, 157, 165, 168, 183–4, 192, 196–7, 201
Swartkrans Cave, Africa, 30
sweet potatoes, 69, 79–80
Switek, B., 227
Syria, 79, 220

Taima-Taima, Venezuela, 63
Tanzania, 26–8, 51, 223
taro, 79–80
Tattersall, I., 229
Teilhard de Chardin, Pierre, 180
Tell Leilan, 84
thermo-energy integrated power system, 241
Thomas, Lewis, 180–1, 244
The Lives of a Cell, 180–1, 244
Thorne, A. et al., 233
three migration hypothesis, 62
Tibetan plateau, 140
Tiwanaka, 92
Tlingit, 58–9, 69, 233
Toca do Boqueirão da Pedra Furada, Brazil, 63
tolerant attitude, 170, 179, 186
tomol, 68
tool-making, 24, 220, 228
Topper, South Carolina, 63
transitional forms, 25
Turkey, 79–80, 85
Turner syndrome, 124
Turney, C.S.M., and Brown, H., 236–7

United Nations Environment Programme, 82, 203, 245
United States of America, iv, 64, 120, 123, 135, 163, 169, 229
Upper Palaeolithic, 37, 55, 57, 210, 212, 223, 226
Ushki Lake, Siberia, 62–3

Van den Bergh, G.D. et al., 232
Van Vark, G.N. et al., 235
Veron, John, 242
volcanic eruptions, 4–5, 12, 43, 54, 76–7, 84, 126
volcanic winter, 54, 84
Vostok, Antarctica, 54, 228, 232

Walker, Alan, 27
Wallace, Alfred Russel, 108
Warren, R., 143, 240
Washington, 64, 66, 69, 220, 232–3, 245
water stress, 137–8, 140, 143
Watters, E., 239
weather, 3, 16, 86, 95, 101, 103, 212, 214, 235, 237
Weaver, A.J., Eby, M. et al., 143
Weaver, A.J., Saenko, O.A. et al., 240
Weaver, Andrew, xiii, 113
Weismann, August, 121
West Antarctic ice sheet, 139, 240
West, F.H., 215
wheat, 86. *See also* emmer and einkorn wheat
White, Tim, 27
Williams, George C., 118
wolves, 40, 77
Wood, B., and Collard, M., 229
Worm, Boris, 182
Wrangel Island, Siberia, 59
Wrangell, Alaska, 69

yams, 79–80
Y-chromosomes, 67, 226
Yenissey River Basin, Siberia, 67
Younger Dryas, 11, 81–2, 86
Yucatan Peninsula, 5, 88–9, 91
Yukon, 59, 64, 211, 233

Zhu, R.X. et al., 229
Zuboff, Robert, 233